Livestock Production and Climate Change

Livestock Production and Climate Change

Editor

Raosaheb Desai

scitus
academics

Livestock Production and Climate Change

Edited by **Raosaheb Desai**

ISBN: 978-1-68117-076-3

Library of Congress Control Number: 2015935474

SCITUS Academics LLC,
www.scitusacademics.com
Box No. 4766, 616 Corporate Way,
Suite 2, Valley Cottage,
NY 10989

Printed in United States of America on Acid Free Paper ∞

Contents

Preface

Climate change is transforming the planet's ecosystems and threatening the well-being of current and future generations. To "hold the increase in global temperature below 2 degrees Celsius" and avoid "dangerous" climate change,2 deep cuts in global emissions are urgently required. The global livestock sector contributes a significant share to anthropogenic GHG emissions, but it can also deliver a significant share of the necessary mitigation effort. Concerted and collective action from all sector stakeholders is urgently required to ensure that existing and promising mitigation strategies are implemented. The need to reduce the sector's emissions and its environmental footprint has indeed become ever more pressing in view of its continuing expansion to ensure food security and feed a growing, richer and more urbanized world population. Technologies and practices that help reduce emissions exist but are not widely used. Their adoption and use by the bulk of the world's producers can result in significant reductions in emissions. Emission intensities (emissions per unit of animal product) vary greatly between production units, even within similar production systems. Different farming practices and supply chain management explain this variability. Within the gap between the production units with the lowest emission intensities and those with the highest emission intensities, lies an important potential for mitigation. A 30 percent reduction of GHG emissions would be possible, for example, if producers in a given system, region and climate adopted the technologies and practice currently used by the 10 percent of producers with the lowest emission intensity.

Editor

Chapter 1

The Influence of Climate, Soil and Pasture Type on Productivity and Greenhouse Gas Emissions Intensity of Modeled Beef Cow-Calf Grazing Systems in Southern Australia

Matthew J. Bell, Brendan R. Cullen, and Richard J. Eckard

Melbourne School of Land and Environment, University of Melbourne, VIC 3010, Australia;

ABSTRACT

A biophysical whole farm system model was used to simulate the interaction between the historical climate, soil and pasture type at sites in southern Australia and assess the balance between productivity and greenhouse gas emissions (expressed in carbon dioxide equivalents, CO_2-eq.) intensity of beef cow-calf grazing systems. Four sites were chosen to represent a range of climatic zones, soil and pasture types. Poorer feed quality and supply limited the annual carrying capacity of the kikuyu pasture compared to phalaris pastures, with an average long-term carrying capacity across sites estimated to be 0.6 to 0.9 cows/ha. A relative reduction in level of feed intake to productivity of calf live weight/ha at weaning by feeding supplementary feed reduced the average CO_2-eq. emissions/kg calf live weight at weaning of cows on the kikuyu pasture (18.4 and 18.9 kg/kg with and without supplementation, respectively), whereas at the other sites studied an increase in intake level to productivity and emission intensity was seen (between 10.4 to 12.5 kg/kg without and with supplementary feed, respectively). Enteric fermentationand nitrous oxide emissions from denitrification were the main sources of annual variability in emissions intensity, particularly at the lower rainfall sites. Emissions per unit product of low input systems can be minimized by efficient utilization of pasture to maximize the annual turnoff of weaned calves and diluting resource input per unit product.

INTRODUCTION

Livestock production systems and the agricultural industries in general face challenges to meet the global demand for food, whilst also abating greenhouse gas (GHG) emissions. Livestock production is one of the major contributors to GHG production globally [1,2]. In Australia, agriculture is a major source of methane (CH_4) and nitrous oxide (N_2O) emissions, which account for 58.9% and 85.9%, respectively, of the net national emissions of these GHGs [3].

Livestock grazing systems in southern Australia are reliant on pasture as a low-cost source of energy and nutrients. Typically, beef cattle grazing systems have low resource input; whereby nitrogen (N) inputs largely come from legume fixation and supplementary feed (in the form of grass silage, hay or grain) if used to fill gaps in pasture supply. Southern Australia has variable climate and pasture systems that need to be managed to cope with inter-annual variability [4]. Grazing systems in this region are largely dependent on the productivity of perennial grass and clover pastures. Climate, soil type and pasture species influence the quality, quantity and reliability of the pasture produced [5]. The productivity of grazing systems, such as live weight gain in cattle, is strongly related to the availability of newly-grown and digestible plant material [6], which can be influenced by pasture management. A winter-active and drought tolerant C3 pasture species such as phalaris (*Phalaris aquatica*) is common in Mediterranean climatic zones in Australia. In comparison, less digestible C4 subtropical grasses like kikuyu (*Pennisetum clandestinum*) are also commonly found in coastal areas of Australia.

Biophysical models can simulate the interactions between the climate, soil properties and grazing animals to predict the pasture produced at a given location. Process-orientated models have been shown [7,8] to be important when comparing regional differences in GHG emissions, particularly when predicting N_2O emissions from soils under different land use. The authors of both studies suggested that for national inventory calculations, more accurate emission factors for soil processes could be obtained based on models that relate soil processes dynamically. One such model is the Sustainable Grazing Systems (SGS) Pasture Model used in this study, which can simulate carbon (C) and N fluxes within the soil. There are few published studies [9,10] that have investigated the influence of geographic location on the relationship between animal productivity, carrying capacity of the system and associated greenhouse gas emission intensity.

This study tests the hypothesis that increased productivity can minimize the GHG emissions intensity of beef cow-calf grazing systems and explores the influence of different climate, soil and pasture types in southern Australia on predictions of (1) the productivity, represented by carrying capacity and the amount of calf live weight at weaning per hectare and (2) the greenhouse gas emissions intensity per hectare and per kg calf live weight at weaning from beef cow-calf grazing systems with and without supplementary feed given. The SGS Model was used to simulate these interactions between pasture species, climate, soil properties and the grazing animal.

EXPERIMENTAL SECTION

Sites Simulated

The SGS Program, which investigated various aspects of grassland productivity and sustainability in southern Australia between the years 1996 to 2001 [11], provided data for the development and testing of the model [12,13]. Four sites were chosen (Table 1) to represent grazing systems with different climatic zones and soil types, which were: Albany [14], Dookie (for this location the SGS site at Ruffy was used for comparison [15]), Vasey [16] and Wagga Wagga [4,13]. Regionally specific dryland (*i.e.*, rainfed) perennial pasture systems were modelled at each site simulated.

Table 1: Sites studied, their location, climate, pasture species and soil type [17] and observed average daily minimum and maximum temperatures and annual rainfall (from the years 1971 to 2000)

Site	Location	Lat., Long.	Climate	Pasture species	Soil type	Temp Min (°C)	Temp Max (°C)	Rainfall (mm)
Albany	SW Western Australia	−34.90, 117.80	Temperate	Kikuyu (Pennisetum clandestinum), subterranean clover (Trifolium subterranean)	Petroferric brown sodosol	10.7	20.5	780
Dookie	N Victoria	−36.37, 145.70	Mediterranean	Phalaris (Phalaris aquatic), subterranean clover, annual ryegrass (Lolium rigidum)	Vertic calic red chromosol	8.1	20.3	576
Vasey	SW Victoria	−38.25, 145.93	Mediterranean	Phalaris, subterranean clover	Yellow sodosol	7.4	19.1	624
Wagga Wagga	S New South Wales	−41.08, 145.77	Mediterranean	Phalaris, subterranean clover, annual ryegrass	Red chromosol/ leptic tenosol	9.6	22.2	549

Whole System Model

A dynamic model incorporating climate data, soil properties, pasture species, livestock and management was used to describe the whole farm system on a daily time step. The SGS Model (documentation and model available at http://www.imj.com.au) has been shown to adequately simulate pasture-based systems with different climates, soil types and pasture species [13,18,19,20]. These studies have tested the predictive ability of the pasture module, which is sensitive to changes in the environment. The productivity of the grazing animal is dependent on the user's definition of the energy requirement of a single animal, as described below, and the energy supplied by feed.

The climate data inputs include minimum and maximum temperature (°C), rainfall (mm), solar radiation (MJ/m^2) and vapor pressure (kPa). Simulations used historical climate data from the SILO database [21] from the years 1961 to 2000 (Table 1).

Cow and Calf Grazing Systems at Each Site

The SGS Model was used to predict the long-term carrying capacity of the pasture base, as these data were not available from the sites simulated. The model does not allow adjustment of cows with offspring on a daily time-step, therefore the carrying capacity was estimated by adjusting the number of steers grazing a one hectare paddock on a daily time-step, which in this case the pasture biomass was grazed to 0.5 t DM ha^{-1} for the phalaris pasture and 1 t DM ha^{-1} for the kikuyu pasture when pasture was available and then animals were removed. The kikuyu pasture was grazed less intensively due to its rhizomatous and stoloniferous growth [22]. Steers were assumed to maintain a 500 kg live weight and have a maintenance energy requirement of 67.5 MJ/d ($0.58 \times 500^{0.75}$ MJ d^{-1} with an additional 10% added for energy expended by activity). The adjusted stocking rate depended on pasture production and its quality as shown in Figure 1. The average number of steers/ha was converted to cows/ha for 30 year simulation runs by assuming a cow with a single calf (referred to as cow + calf from hereon) has a metabolizable energy (ME) requirement of 1.5 times that of a steer with a 500 kg live weight [23].

Throughout the year cows grazed four paddocks in a grazing system that optimized pasture intake/ha at each site. Relative comparisons could therefore be made between the sites studied, as the management of the grazing systems simulated at each site was the same. The timing of calving (the first day of the month) was set to coincide with when the cows annual pasture intake/ha was greatest, whilst also minimizing the need for supplementary feed. Cows calved at the start of July at each site (Figure 1), with an assumed 100% calving percentage and no calf mortality. The maximum growth rate of the calf was set at 0.6 kg d^{-1} from birth until weaning at 120 days of age, the point at which the calf was removed from the system. The target live weight of a calf at weaning was 100 kg. Cows and their calves were allowed to move between the four paddocks so that the available pasture biomass was maintained between 0.5 to 2.5 t DM ha^{-1} for a phalaris pasture and 1 to 2.5 t DM ha^{-1} for the kikuyu pasture. Any surplus pasture was cut to maintain a vegetative and palatable pasture for the grazing animal, at a time when pasture growth can be high, particularly in September to November (Figure 1). When the pasture biomass reached 2.5 t DM ha^{-1} it was assigned for cutting, which occurred at 3.5 t DM ha^{-1}.

The animal module describes the animal's energy requirements for maintenance (defined as $0.58 \times$ live weight$^{0.75}$ MJ d^{-1}), feeding activity, pregnancy, lactation and live weight change (see documentation at http://www.imj.com.au for a full description). The model assumes that the animal will try to eat enough feed to meet its ME requirement, with ME intake constrained by the ME content of available feed and a constraint on the animal's maximum DM intake of 16.9 kg [24]. The ME in this study was for the cow + calf, with any limitation on calf growth being due to a restriction on ME consumption.

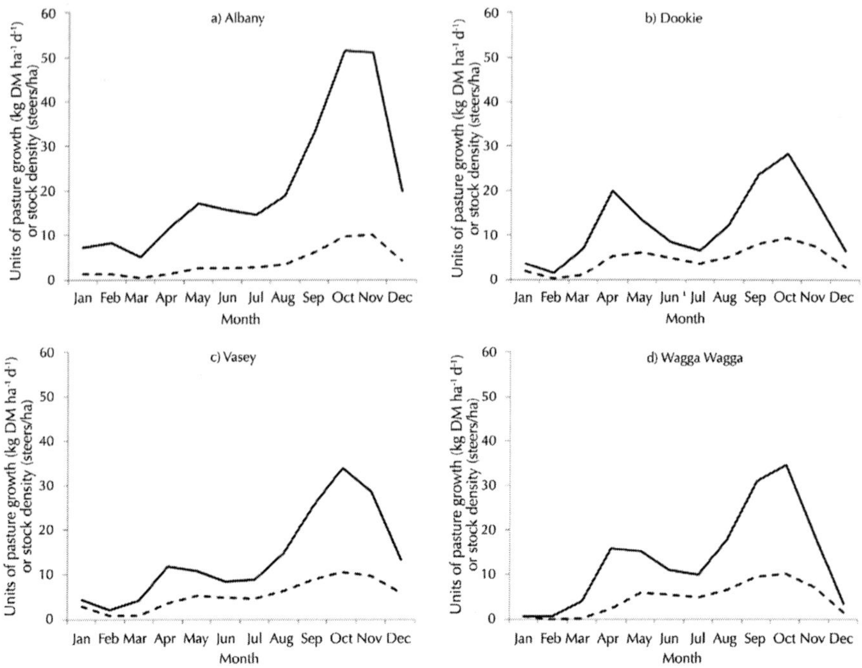

Figure 1: Predicted average daily pasture growth (—; kg DM ha^{-1} d^{-1}) and stock density (- -; steers/ha) from January to December for the years of 1971 to 2000 at (**a**) Albany, (**b**) Dookie, (**c**) Vasey and (**d**) Wagga Wagga.

Simulations were run with and without supplementary feed being given to the cow + calf. Therefore the diet consisted of either solely pasture or pasture and supplement. When there was no feed available, the cow and its calf would be removed from the system. When the diet was supplemented, grass silage and/or bought-in low quality grain were assumed to be fed to meet the animal's ME requirement and maintain the live weight of the cow above 450 kg and lactation requirement. The model does not predict the quality of grass silage produced on the farm, which is being fed back to the animal. Therefore an assumption had to be made about the quality of the supplementary feed, which was fixed at 10.5 MJ kg DM^{-1} for ME content and 2.5% nitrogen for sites with a phalaris pasture system and an ME content of 10 MJ kg DM^{-1} and 2.0% nitrogen for the site with a kikuyu pasture system. Any surplus grass was cut and removed or made available for feed in the form of grass silage; with 10% wastage assumed during the cutting process. Due to the modeling approach used, the grass silage was only available as feed during the twelve months after being harvested (after which time it was removed from the system), which is a limitation of the model.

Nitrogen inputs within the grazing system came from plant residues, dung, urine and nitrogen fixation by legumes, with no synthetic fertilizer used. The model is based on the assumption that nitrogen, water and temperature can all limit pasture growth. The simulations assumed sufficiency of nutrients other than nitrogen and water.

Grazing System Greenhouse Gas Emissions

The CO_2-eq. emissions associated with each site for the cow + calf over a year were predicted. Total system CO_2-eq. emissions were calculated up to the farm-gate using the same approach to account for total production system emissions as that used by [25] when assessing dairy systems, except the present study did not include the GHG emissions produced by replacement animals. Emissions from the grazing system were assessed by CO_2-eq. emissions per hectare and per kg of total calf live weight at weaning (the functional units as defined by [26]). The CO_2-eq. emissions were calculated using conversion factors from CH_4 and N_2O to CO_2-eq. emissions of 21 and 310 (for a 100 year time horizon) respectively [3]. The sources of CO_2-eq. emissions were from enteric fermentation, manure and soil (directly through denitrification and indirectly through leaching and volatilization) and from feed production and use *i.e.*, for pasture, grass silage and bought-in grain. The methods for calculating GHG emissions are given in the.

Data Analysis

The first 10 years of each simulation were discarded to allow the nutrient dynamics within the model to stabilize, leaving a data set from 1971 to 2000 for analysis. Predictions of productivity and emission intensity of sites were obtained with climate, soil type and pasture species as an input to the model. The productivity of each grazing system was assessed by predicting the annual carrying capacity of the pasture system (cow + calf/ha), average annual intake of pasture (t DM ha^{-1}) and the average calf live weight at weaning (kg ha^{-1}). The productivity of the system was compared to the CO_2-eq. emissions per ha and per kg calf live weight at weaning. The CO_2-eq. emissions per kg calf live weight were classified by emissions source: enteric CH_4, manure and soil GHG, and GHG from feed production (for pasture, grass silage and bought-in grain).

RESULTS

Productivity

The average long-term carrying capacity of the pasture at Albany, Dookie, Vasey and Wagga Wagga was predicted to be cows per hectare of 0.6, 0.8, 0.9 and 0.8, respectively. Based on the long-term carrying capacity estimated by the model, simulations were set up to assess the variability in production and GHG emissions of beef cow-calf grazing systems. At each site studied, the predicted peak of average pasture growth during the years 1971 to 2000 was in the spring (September to November) when cows were with their calves (Figure 1).

The peak of average spring pasture growth was notably greater at Albany with the kikuyu pasture than at other sites studied. Albany also has a temperate climate and is of higher annual rainfall (Table 1). All the grazing systems simulated were predicted to maintain the cow's daily live weight above 450 kg and ME requirement between 81 to 85 MJ/d (Table 2).

Table 2: Mean (s.d) annual predicted cow live weight and their offspring, total metabolizable energy (ME) requirement, the quality of the diet consumed (ME content and digestibility) and surplus dry matter (DM) cut for beef cow-calf [1] grazing systems at Albany, Dookie, Vasey and Wagga Wagga between the years 1971 to 2000 on pasture or a pasture and supplementary feed [2]diet

Site	Units	Pasture	Pasture + supplementary feed
Albany	Kikuyu/subclover pasture and stocking rate 0.6 cows/ha		
Cow live weight	kg/hd/d	474 (11)	474 (12)

Calf live weight at weaning	kg/hd	96.2 (7.2)	100 (0.2)
ME required	MJ/hd/d	85.3 (1.1)	85.9 (0.5)
Pasture intake	t DM/ha	1.9 (0.2)	1.9 (0.1)
Supplement intake	t DM/ha	0	0.04 (0.05)
ME content	MJ/kg DM	9.6 (0.1)	9.7 (0.1)
DM digestibility	%	60.2 (0.8)	60.3 (0.8)
Pasture cut	t DM/ha	0	1.5 (0.7)
Dookie	Phalaris/subclover/annual ryegrass pasture and stocking rate 0.8 cows/ha		
Cow live weight	kg/hd/d	455 (3)	457 (3)
Calf live weight at weaning	kg/hd	62.5 (18)	100 (0.2)
ME required	MJ/hd/d	81.3 (2.1)	86.5 (0.2)
Pasture intake	t DM/ha	0.8 (0.5)	1.0 (0.4)
Supplement intake	t DM/ha	0	0.5 (0.2)
ME content	MJ/kg DM	9.8 (0.3)	10.2 (0.1)
DM digestibility	%	60.9 (1.7)	63.6 (0.6)
Pasture cut	t DM/ha	0	0.3 (0.7)
Vasey	Phalaris/subclover pasture and stocking rate 0.9 cows/ha		
Cow live weight	kg/hd/d	456 (1)	457 (1)
Calf live weight at weaning	kg/hd	73.4 (10)	100 (0.2)
ME required	MJ/hd/d	82.5 (1.3)	86.3 (0.1)
Pasture intake	t DM/ha	1.2 (0.3)	1.5 (0.3)

Supplement intake	t DM/ha	0	0.5 (0.1)
ME content	MJ/kg DM	9.9 (0.2)	10.2 (0.1)
DM digestibility	%	61.9 (1.0)	63.5 (0.5)
Pasture cut	t DM/ha	0	0.6 (0.6)
Wagga Wagga	Phalaris/subclover/annual ryegrass pasture and stocking rate 0.8 cows/ha		
Cow live weight	kg/hd/d	456 (2)	457 (3)
Calf live weight at weaning	kg/hd	74.9 (16)	100 (0.2)
ME required	MJ/hd/d	83.1 (1.8)	86.5 (0.2)
Pasture intake	t DM/ha	0.9 (0.4)	1.1 (0.4)
Supplement intake	t DM/ha	0	0.3 (0.2)
ME content	MJ/kg DM	10.0 (0.3)	10.2 (0.1)
DM digestibility	%	62.4 (1.6)	64.9 (0.7)
Pasture cut	t DM/ha	0	0.6 (0.8)

[1] A cow over a full year plus its calf from birth until weaning at 120 days of age.

[2] Grass silage and/or grain

On average, the total feed intake for a cow + calf per year ranged from 0.8 to 1.9 t DM/ha on pasture and 1.4 to 2.0 t DM/ha on a pasture and supplementary feed diet (Table 2). The use of supplementary feed meant that the target live weight of a calf at weaning of 100 kg was achieved at all sites. However, with no supplementary feed given, the average live weight of a calf ranged from 63 kg at the lower rainfall site of Dookie compared to 96 kg at Albany, which has a higher rainfall and kikuyu pasture.

Table 2 shows that the average annual DM digestibility and ME content of the diet at Albany (which averaged 60% and 9.7 MJ/kg DM respectively across diets) was poorer than at other sites (ranging from 61 to 65% and 9.8 to 10.2 MJ/kg DM, respectively). Surplus pasture was cut to help maintain the availability of new and digestible plant material.

Greenhouse Gas Emissions and Productivity

Enteric CH_4 was the main source of emissions from all grazing systems studied (Table 3). The proportion of GEI lost as enteric CH_4 was predicted to be lower for the kikuyu pasture systems, averaging 7.4% compared to 7.8% for animals on the phalaris based pasture at the other sites. There were small differences between sites and pasture systems in their average CH_4 produced per kg of digestible dry matter (DDM), which was lowest at Albany (averaging 40.6 g/kg DDM) and slightly higher at the lower rainfall sites of Dookie and Wagga Wagga (averaging 41.7 g/kg DDM). At each site, the main source of N_2O emissions per hectare was by denitrification, followed by volatilization and then leaching. The emissions of N_2O by denitrification varied considerably at the lower rainfall sites of Dookie and Wagga Wagga compared to the higher rainfall sites of Albany and Vasey.

Table 3: Annual mean (s.d) carbon dioxide (CO_2-eq.) equivalent emissions per hectare from enteric methane (CH_4; expressed as percentage of gross energy intake (GEI) and per digestible dry matter (DDM)), manure CH_4, nitrous oxide (N_2O) emissions by leaching, volatilization and denitrification and kilograms of nitrogen fixed by legumes predicted for beef cow-calf [1] grazing system at Albany, Dookie, Vasey and Wagga Wagga between the years 1971 to 2000 on pasture or a pasture and supplementary feed [2]diet

Site	Units	Pasture	Pasture + supplementary feed
Albany	Kikuyu/subclover pasture and Stocking rate 0.6 cows/ha		
Enteric CH4	kg CO2-eq./ha	926 (68)	944 (51)
Enteric CH4	% GEI	7.4 (0.1)	7.4 (0.1)
Enteric CH4	g CH4/kg DDM	40.6 (0.3)	40.6 (0.3)
Manure CH4	kg CO2-eq./ha	29.7 (2.7)	30.2 (2.2)
Leaching	kg CO2-eq./ha	5.3 (3.1)	5.2 (2.9)
Volatilisation	kg CO2-eq./ha	12.6 (2.2)	12.8 (2.5)

Denitrification	kg CO2-eq./ha	99.3 (54.7)	100 (56.3)
Legume fixation	kg N/ha	58.4 (18.6)	64.7 (17.2)
Total CO2-eq.	t/ha	1.1 (0.1)	1.1 (0.1)
Dookie	Phalaris/subclover/annual ryegrass pasture and Stocking rate 0.8 cows/ha		
Enteric CH4	kg CO2-eq./ha	427 (260)	825 (137)
Enteric CH4	% GEI	7.8 (0.1)	7.8 (0.1)
Enteric CH4	g CH4/kg DDM	42.4 (1.0)	41.2 (0.6)
Manure CH4	kg CO2-eq./ha	12.6 (7.4)	22.9 (4.3)
Leaching	kg CO2-eq./ha	0.1 (0.3)	0.2 (0.3)
Volatilisation	kg CO2-eq./ha	5.3 (3.9)	7.9 (2.7)
Denitrification	kg CO2-eq./ha	93.9 (110.9)	100 (112)
Legume fixation	kg N/ha	16.4 (6.0)	16.9 (6.5)
Total CO2-eq.	t/ha	0.6 (0.4)	0.9 (0.2)
Vasey	Phalaris/subclover pasture and Stocking rate 0.9 cows/ha		
Enteric CH4	kg CO2-eq./ha	645 (148)	1055 (98)
Enteric CH4	% GEI	7.8 (0.1)	7.8 (0.1)
Enteric CH4	g CH4/kg DDM	41.9 (0.5)	40.8 (0.3)
Manure CH4	kg CO2-eq./ha	18.7 (4.2)	29.7 (3.2)
Leaching	kg CO2-eq./ha	1.6 (1.8)	1.8 (2.0)
Volatilisation	kg CO2-eq./ha	8.8 (2.8)	12.1 (3.3)
Denitrification	kg CO2-eq./ha	42.6 (23.0)	47.0 (23.9)
Legume fixation	kg N/ha	26.2 (7.8)	29.8 (7.2)
Total CO2-eq.	t/ha	0.7 (0.2)	1.1 (0.1)
Wagga Wagga	Phalaris/subclover/annual ryegrass pasture and Stocking rate 0.8 cows/ha		
Enteric CH4	kg CO2-eq./ha	501 (228)	807 (104)
Enteric CH4	% GEI	7.9 (0.1)	7.9 (0.1)
Enteric CH4	g CH4/kg DDM	42.0 (0.9)	41.2 (0.6)
Manure CH4	kg CO2-eq./ha	14.1 (6.2)	22.1 (3.2)

Leaching	kg CO2-eq./ha	0.001 (0.003)	0.002 (0.004)
Volatilisation	kg CO2-eq./ha	6.8 (3.6)	9.2 (2.9)
Denitrification	kg CO2-eq./ha	124.3 (97.1)	136 (106)
Legume fixation	kg N/ha	18.0 (6.0)	19.3 (6.6)
Total CO2-eq.	t/ha	0.7 (0.3)	1.0 (0.2)

[1]A cow over a full year plus its calf from birth until weaning at 120 days of age.

[2]Grass silage and/or grain.

The average annual CO_2-eq. emissions/ha were 1.1 ± 0.1 t CO_2-eq. at Albany, 0.6 ± 0.4 and 0.9 ± 0.2 t CO_2-eq. at Dookie, 0.7 ± 0.2 and 1.1 ± 0.1 t CO_2-eq. at Vasey, and 0.7 ± 0.3 and 1.0 ± 0.2 t CO_2-eq. at Wagga Wagga (Table 3) on pasture or pasture with supplementary feed, respectively. Notably, the CO_2-eq. emissions/ha increased with supplementary feed being given at Dookie, Vasey and Wagga Wagga as the average live weight of a calf at weaning increased (Table 2).

Enteric CH_4 contributed between 0.75 to 0.93, manure and soil between 0.03 to 0.21, and feed production from 0.02 to 0.06 of total CO_2-eq. emissions per kg calf live weight at weaning across pasture systems (Figure 2). The low stocking rate of cows at Albany compared to their amount of feed consumed contributed to its higher annual enteric CH_4 emissions per kg calf live weight and CO_2-eq. emissions per kg calf live weight at weaning. Annual CO_2-eq. emissions per kg calf live weight at weaning averaged 18.9 ± 1.1 and 18.4 ± 1.5 kg CO_2-eq. kg^{-1} at Albany, 10.4 ± 4.3 and 11.6 ± 3.0 kg CO_2-eq. kg^{-1} at Dookie, 11.0 ± 1.8 and 12.5 ± 1.4 kg CO_2-eq. kg^{-1} at Vasey and 10.7 ± 3.8 and 11.9 ± 2.5 kg CO_2-eq. kg^{-1} at Wagga Wagga on pasture or pasture with supplementary feed, respectively.

As was seen for CO_2-eq. emissions/ha, the CO_2-eq. emissions per kg calf live weight increased (Figure 2 and Figure 3) with supplementary feed being given at Dookie, Vasey and Wagga Wagga as the average live weight of a calf at weaning increased (Table 2) and turnoff of calf live weight at each site increased resulting in more enteric CH_4 emissions per kg calf live weight. Figure 3 shows the CO_2-eq. emissions per kg calf live weight at Albany showed a slight decline with an increase in productivity with supplementary feed being fed.

The use of supplementary feed increased the turnoff of calf live weight at weaning across sites. Due to its higher capacity to carry stock, the phalaris/subclover grazing system at Vasey had the highest predicted total calf live weight at weaning of 66 and 90 kg/ha compared to 60 and 80 kg/ha at Wagga Wagga, 50 and 80 kg/ha at Dookie and 58 and 60 kg/ha at Albany for the pasture or pasture and supplementary feed system, respectively.

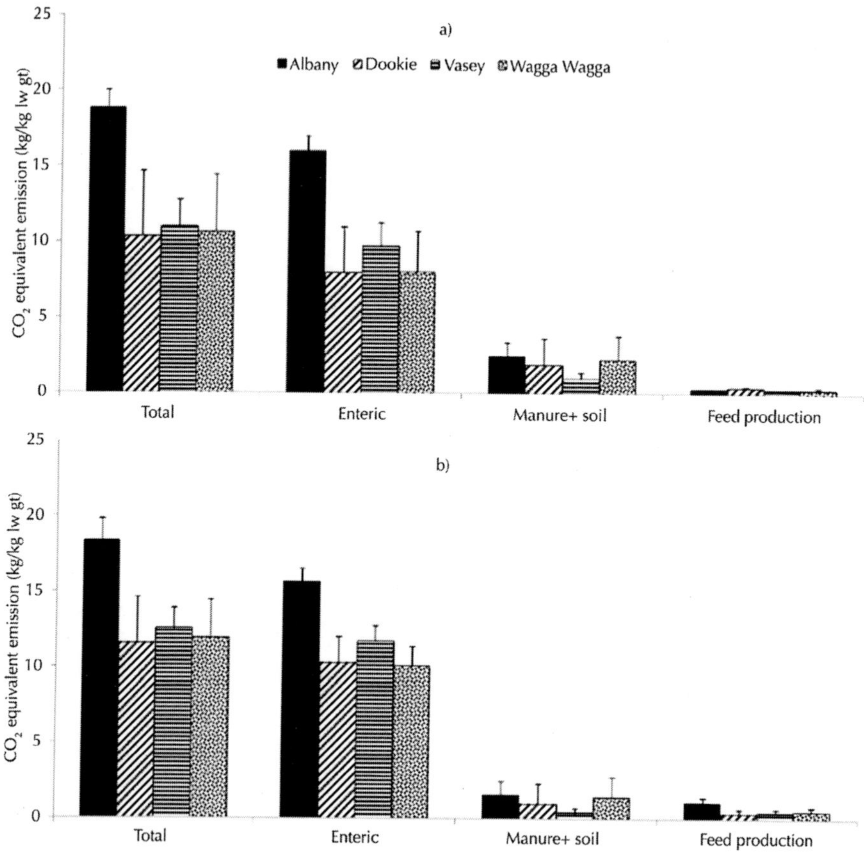

Figure 2: Predicted average annual carbon dioxide (kg CO_2-eq.) equivalent emissions per kg calf live weight at weaning (kg lwgt) from all sources (total), enteric fermentation, manure and soil, feed production (pasture, forage and bought-in grain) at Albany, Dookie, Vasey and Wagga Wagga for the years 1971 to 2000 for cows fed a) pasture or b) pasture and supplementary feed. Vertical bars indicate standard deviation for 30-year model runs.

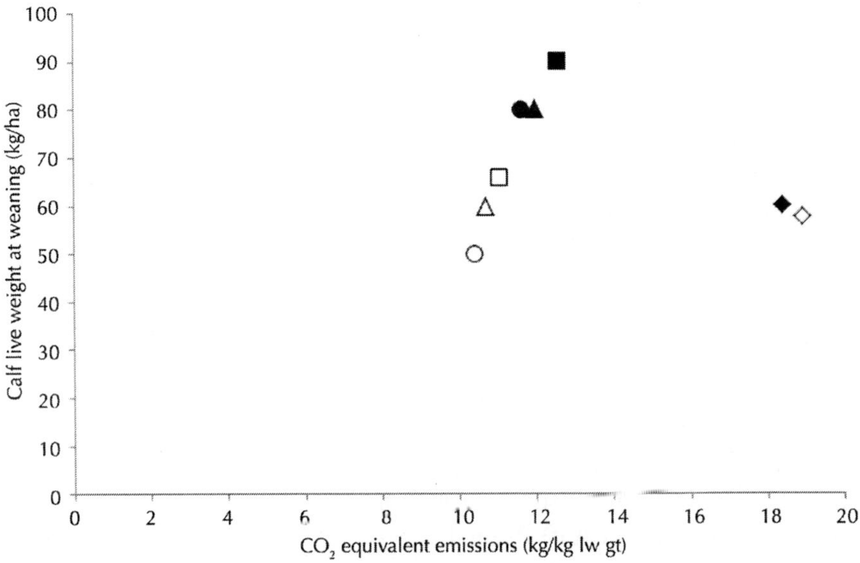

Figure 3: Predicted average annual carbon dioxide (kg CO_2-eq.) equivalent emissions per kg calf live weight at weaning (kg lwgt) and turnoff of calf live weight at weaning (kg) per hectare at Albany (diamond), Dookie (circle), Vasey (square) and Wagga Wagga (triangle) for the years 1971 to 2000 for cows on pasture (white symbols) or pasture and supplementary feed (black symbols).

DISCUSSION

Productivity

Traditionally beef cow-calf systems in southern Australia use minimal inputs and rely on mixed pastures as a low-cost source of energy and nutrients, rather than bought-in feeds. The long-term carrying capacities predicted in this study for Albany, Dookie, Vasey and Wagga Wagga of 0.6, 0.8, 0.9 and 0.8 cows per hectare are comparable to those reported for the Albany site [22] of 0.6 cows per hectare and average values reported [12] for phalaris/subclover pastures at the sites of Vasey and Maindample (in Victoria) of 0.8 to 0.9 cows per hectare. This provided some confidence that the energy requirements of the grazing animal were being appropriately defined in the model.

Of the sites studied, the most productive grazing system was a phalaris/subclover pasture at Vasey. Vasey was predicted to carry a stocking rate of 0.9 cows/ha and produce an average of 66 kg/ha on pasture and 90 kg/ha of calf live weight at weaning on pasture and supplementary feed, compared to 50 to 80 kg/ha across the other sites. The lowest stocking rate modelled was at Albany with 0.6 cows/ha for the kikuyu pasture system. A kikuyu pasture has been found to be productive if the pasture base has a high proportion of legume such as subclover [22]. The kikuyu/subclover pasture was able to fix on average 62 kg N ha^{-1} yr^{-1}, which was more than at other sites (ranging from 17 to 28 kg N ha^{-1} yr^{-1} on average). In the study by [22], 40% of the farm was planted with a kikuyu/subclover pasture and the remainder to annual species, which seems appropriate management to utilize the high pasture production of the C4 pasture during the spring and early summer months (September to December in Figure 1) by adjusting livestock numbers to prevent a build-up of less digestible pasture—as occurred in this study with high amounts of pasture cut compared to the phalaris pastures modeled (Table 2). The benefit of kikuyu is that during the warmer months, its deep and adventitious root structure makes it efficient at extracting soil nutrients and water [27,28] and to be productive on deep and freely draining coastal soils such as at Albany [22,29]. Therefore this pasture system can be a useful source of green feed in this location but requires appropriate management to be utilized effectively. A C4 pasture species like kikuyu is better suited to warmer sub-tropical latitudes [30], but may become a more dominant pasture species in the future with warming [10].

Greenhouse Gas Emissions and Productivity

Management, feeding and genetics can all potentially play an important role in reducing system enteric CH_4 emissions, particularly per unit product [31,32,33]. For all systems modeled, enteric CH_4 was predicted to be the main source of emissions (0.75 to 0.93 of total system emissions). The annual emissions produced by a cow and its calf in this study (below 75 kg/year at all sites; Table 3) were comparable to published measurements for grazing systems [33]. Given concerns about the quality of C4 dominant pastures compared to those with C3 grasses (such as phalaris) and its impact on enteric

CH_4 emissions [5], this study suggests that a kikuyu pasture system may not necessarily produce more enteric CH_4 as a percentage of GEI or per kg DDM on average over a year than the C3 pastures studied (7.4% and 40.6 g/kg DDM compared to 7.8% and 41.7 g/kg DDM on average, respectively). The lower enteric CH_4 lost as a percentage of GEI from a kikuyu grazing system, compared to the C3 systems, could be attributed to the higher level of feed intake per animal. On average the ME content and DM digestibility of the kikuyu pasture system (9.7 MJ/kg DM and 60% respectively) was poorer than the phalaris diets (averaging 9.8 to 10.2 MJ/kg DM and 61 to 65%, respectively). The losses of enteric CH_4 in this study from a kikuyu pasture were higher than values found by [34] studying grazing dairy cows, who reported measured values for losses of enteric CH_4 of 7.1% of GEI and 33.8 g/kg DDM. Given the variability in pasture quality, the authors [34] found these values were not that different from a summer perennial ryegrass/white clover pasture and losses of enteric CH_4 appeared to reduce from the kikuyu pasture when N fertilizer was applied, which improved its DM digestibility. While the model used in this study by [35] to predict enteric methane losses is the accepted method in Australian GHG Inventory, there are concerns that it may over-predict losses from Australian forages [36]. This does not affect comparisons between sites but more the relative contribution of enteric methane to total systems emissions.

Climate, soil type and management of the pasture species influence the availability of young and digestible plant material, which in turn strongly affects the productivity of the system [5,6]. A study on sheep grazing systems [37] found that management structures and genetic improvements that give the highest sustainable economic yield can also give the lowest enteric CH_4 emissions intensity per unit product. The average enteric CH_4 emissions per unit product from the kikuyu pasture system were notably high at 16.1 kg CO_2-eq. emissions per kg from cows on solely pasture, but declined to 15.7 kg CO_2-eq. emissions per kg with the addition of supplementary feed, due to the relative level of feed intake to productivity of calf live weight/ha at weaning. At the other sites studied, the average enteric CH_4 emissions per unit product increased from cows fed pasture and supplement compared to solely grazing pasture, due to an increase in feed intake relative to productivity. In this study and similar to [37], the timing of calving was chosen to coincide with the onset of winter pasture growth, so that pasture supply met the increasing demand from the cow and its calf as well as being of high digestibility and optimizing pasture use (their suggested economic optimum time for parturition). The management of livestock systems can vary considerably, but as found in other studies [37], the efficient use of herbage consumed whilst maintaining optimum animal numbers is important for minimizing grazing system emissions. In this study the grazing system was modeled to provide a digestible pasture by grazing four paddocks and cutting any surplus pasture. Rather than this imposed management on the grazing system, the sites with a more poorly drained chromosol soil type (Dookie and Wagga Wagga) and lower rainfall may benefit from a more flexible grazing management approach e.g., an adjustment in stocking rate. These sites may rely more on supplementary feed to maintain their production of weaned calf or wean calves at a younger age; likely a function of their more variable annual supply of pasture compared to the higher rainfall sites modeled. Developments of the SGS model are required to allow more flexible management options to be assessed. Another limitation of the SGS model is its lack of flexibility in terms of adjusting the number of mature animals with offspring based on the availability of pasture and the sensitivity of the animal to its environment (this is captured more in the growth of the pasture plant). Overall, the use of supplementary feed would contribute a small proportion to total system CO_2-eq. emissions (0.02 to 0.06).

The emissions from the grazing system studied could be higher if additional resources were to be sourced from outside the farm unit. As this study suggests and other studies have shown [38,39], pasture-based livestock systems have the potential to be productive and minimize their GHG emissions intensity per unit product and per hectare by being efficient users of natural resources rather than bought-in farm inputs if possible. In comparison to other beef cow production systems which show a wide range of emission intensity, the annual emission intensity per unit product in this study ranged from 10.4 to 18.9 and 11.6 to 18.4 kg CO_2-eq. emissions kg^{-1} of calf live weight at weaning for pasture or pasture and supplementary feed diets compared to 15.8 to 48.1 kg CO_2-eq. emissions kg^{-1} of live weight at slaughter for a range of systems in the study of [40] (assuming a dressing percentage of 0.53). In the study by [40] comparing the emission intensity per unit product of beef production, the lowest value reported was for an African pastoral system, whereas the higher intensities were from systems that required higher resource inputs. The feeding of supplement in this study to overcome the constraint of pasture supply did not reduce overall GHG emissions per hectare or emissions per unit product. When pasture supply was adequate and supplement was fed, as simulated in the kikuyu grazing system, a reduction in emissions per unit product was observed with a slight improvement in feed quality and increase in production of calf live weight per hectare at weaning. In beef systems, where enteric CH_4 is the predominant source of emissions and feed intake is not constrained, improving the feed quality has been found to reduce emissions per unit product by promoting more rapid growth and reducing the number of days to reach a target live weight [41]. This study suggests that there is regional variation in emission intensity per unit product partly caused by differences in soil emissions, feed quality and utilization, which may not necessarily be represented in other studies that do not use location specific inputs to a process-based model. In the present study, the GHG emissions from replacement animals were not included, which can be a significant source of emissions. Another study [42] on beef cow and calf production systems in the southern Australian region included the emissions associated with replacements for a self-replacing breeding system, which as discussed by the authors, inflated the predicted emissions intensity per kg calf live weight (averaging 45.6 kg CO_2-eq. emissions kg^{-1} of live weight at slaughter) compared to those reported by [42].

At the lower rainfall sites of Dookie and Wagga Wagga, variability in manure and soil emissions per ha and per unit product were largely due to N_2O emissions from denitrification in soil and meant that manure and soil emissions constituted a greater proportion of system emissions on average. The model parameters for simulating nitrogen losses and estimating N_2O emissions have been compared to measured emissions in a previous study by [43] and were found to provide realistic estimates of N_2O loses and behavior. The capacity for denitrification within soil is increased with conditions that increase the availability of carbon, temperatures that increase microbial activity and saturation of the WFPS to create anaerobic conditions [44]. Even though more freely draining soils at Albany and Vasey were a greater source of N_2O leaching losses than at the other sites studied, which was expected [8], losses of N_2O from leaching and N_2O losses from N volatilization were less of a source of N_2O emissions on average compared to denitrification in soil. The differences in N_2O emissions between sites emphasizes the need for a dynamic modeling approach to accounting for N_2O emissions from grazing systems in different regions, as opposed to static Tier 1 and 2 emission factors, and would provide a better reflection of these systems in national inventory estimates.

The study used a process-based model to predict both enteric methane and nitrous oxide emissions. As emissions of these gases can be highly variable and are inherently difficult to both measure and model, there are inherent uncertainties in both the model predictions and the measured data sources. Therefore, while the model predictions presented in this study are within plausible boundaries, the emphasis of this study is more on the relative difference between systems modeled at different sites, rather than the absolute values predicted. Process-based models are well suited to predict these relative differences and allow exploration of system behavior.

CONCLUSIONS

In conclusion, while it may not be economically viable, the model simulations of low input grazing systems indicate that the feeding of supplementary feed in the form of grass silage or grain should improve the productivity, but increase CO_2-eq. emissions per ha. Enteric CH_4 was the main source of CO_2-eq. emissions per ha and per unit product. On average, a reduction in enteric CH_4 emissions per unit product due to the relative level of feed intake to productivity of calf live weight/ha at weaning *i.e.*, a dilution of resource input per unit product, may reduce emission intensity as seen in a supplemented subtropical kikuyu pasture system in this study. By understanding the effect that climatic, edaphic characteristics associated with a particular location can have on GHG emissions, in combination with management decisions, producers could manage their grazing system to maintain low emission intensity per unit product by efficient use of pasture to optimize its carrying capacity.

ACKNOWLEDGMENTS

This work was supported by funding from Dairy Australia, Meat and Livestock Australia and the Australian Government Department of Agriculture, Fisheries and Forestry under its Australia's Farming Future Climate Change Research Program.

REFERENCES

1. Steinfeld, H.; Gerber, P.; Wassenaar, T.; Castel, V.; Rosales, M.; de Haan, C. *Livestock's Long Shadow—Environmental Issues and Options; FAO Report*; FAO: Rome, Italy, 2006.
2. Intergovernmental Panel on Climate Change. *Climate Change Series*; Cambridge University Press: New York, NY, USA, 2007.
3. *The Australian Government Submission to the UN Framework Convention on Climate Change; National Inventory Report 2007—Volume 1*; Department of Climate Change and Energy Efficiency: Canberra, ACT, Australia, 2009.
4. Cullen, B.R.; Johnson, I.R.; Eckard, R.J.; Lodge, G.M.; Walker, R.G.; Rawnsley, R.P.; McCaskill, M.R. Climate change effects on pasture systems in south-eastern Australia. *Crop Pasture Sci.* 2009, *60*, 933–942.
5. Howden, S.M.; Crimp, S.J.; Stokes, C.J. Climate change and Australian livestock systems: Impacts, research and policy issues. *Aust. J. Exp. Agric.* 2008, *48*, 780–788.
6. Ash, A.J.; Prinsen, J.H.; Myles, D.J.; Hendricksen, R.E. Short-term effects of burning native pasture in spring on herbage and animal production in south-east Queensland. *Proc. Austr. Soc. Anim. Prod.* 1982, *14*, 377–380.
7. Boeckx, P.; van Cleemput, O. Estimates of N_2O and CH_4 fluxes from agricultural lands in various regions in Europe. *Nutr. Cycl. Agroecosyst.* 2001, *60*, 35–47.
8. Eckard, R.J.; Cullen, B.R. Impacts of future climate scenarios on N_2O emissions from pasture-based dairy systems in SE Australia. *Anim. Feed Sci. Technol.* 2011, *166-167*, 736–748.

9. Cullen, B.R.; Eckard, R.J. Impacts of future climate scenarios on the balance between productivity and total greenhouse gas emissions from pasture based dairy systems in south-eastern Australia. *Anim. Feed Sci. Technol.* 2011, *166-167*, 721–735.

10. Bell, M.; Eckard, R.; Cullen, B. The effect of future climate scenarios on the balance between productivity and greenhouse gas emissions from sheep grazing systems. *Livest. Sci.* 2012, *147*, 126–138.

11. Mason, W.K.; Lamb, K.; Russell, B. The Sustainable Grazing Systems Program: New solutions for livestock producers. *Aust. J. Exp. Agric.* 2003, *43*, 663–672.

12. Andrew, M.H.; Lodge, G.M. The Sustainable Grazing Systems National Experiment: 1. Introduction and Methods. *J. Exp. Agric.* 2003, *43*, 695–709.

13. Johnson, I.R.; Lodge, G.M.; White, R.E. The Sustainable Grazing Systems Pasture Model: Description, philosophy and application to the SGS National Experiment. *Aust. J. Exp. Agric.* 2003, *43*, 711–728.

14. Sandford, P.; Wang, X.; Greathead, K.D.; Gladman, J.H.; Speijers, J. Impact of Tasmanian blue gum belts and kikuyu-based pasture on sheep production and groundwater recharge in south-western Western Australia. *Aust. J. Exp. Agric.* 2003, *43*, 755–767.

15. Ridley, M.; Christy, B.P.; White, R.E.; McLean, T.; Green, R. North-East Victoria SGS National Experiment site: Water and nutrient losses from grazing systems on contrasting soil types and levels of inputs. *Aust. J. Exp. Agric.* 2003, *43*, 799–815.

16. Chapman, D.F.; McCaskill, M.R.; Quigley, P.E.; Thompson, A.N.; Graham, J.F.; Borg, D.; Lamb, J.; Kearney, G.; Saul, G.R.; Clark, S.G. Effect of grazing method and fertiliser inputs on the productivity and sustainability of phalaris-based pastures in Western Victoria. *Aust. J. Exp. Agric.* 2003, *43*, 785–798.

17. Isbell, R.F. *Australian Soil and Land Survey Handbook: The Australian Soil Classification*; CSIRO Publishing: Collingwood, VIC, Australia, 1996.

18. Cullen, B.R.; Eckard, R.J.; Callow, M.N.; Johnson, I.R.; Chapman, D.F.; Rawnsley, R.P.; Garcia, S.C.; White, T.; Snow, V.O. Simulating pasture growth rates in Australian and New Zealand grazing systems. *Aust. J. Exp. Agric.* 2008, *59*, 761–768.

19. Johnson, I.R.; Chapman, D.F.; Snow, V.O.; Eckard, R.J.; Parsons, A.J.; Lambert, M.G.; Cullen, B.R. DairyMod and EcoMod: Biophysical pastoral simulation models for Australia and New Zealand. *Aust. J. Exp. Agric.* 2008, *48*, 621–631.
20. Lodge, G.M.; Johnson, I.R. Agricultural drought analyses for temperate Australia using a biophysical pasture model. 1. Identifying and characterising drought periods. *Aust. J. Exp. Agric.* 2008, *59*, 1049–1060.
21. Jeffrey, S.J.; Carter, J.O.; Moodie, K.M.; Beswick, A.R. Using spatial interpolation to construct a comprehensive archive of Australian climate data. *Environ. Modell. Softw.* 2001, *16*, 309–330.
22. McDowall, M.M.; Hall, D.J.M.; Johnson, D.A.; Bowyer, J.; Spicer, P. Kikuyu and annual pasture: A characterization of a productive and sustainable beef production system on the South Coast of Western Australia. *Aust. J. Exp. Agric.* 2003, *43*, 769–783.
23. *Drought Feeding and Management of Beef Cattle: A Guide for Farmers and Land Managers*; Department of Primary Industries, Victorian Government: Melbourne, Australia, 2007.
24. Neal, M.; Neal, J.; Fulkerson, W.J. Optimal choice of dairy forages in eastern Australia. *J. Dairy Sci.* 2007, *90*, 3044–3059.
25. Bell, M.J.; Wall, E.; Russell, G.; Simm, G.; Stott, A.W. The effect of improving cow productivity, fertility, and longevity on the global warming potential of dairy systems. *J. Dairy Sci.* 2011, *94*, 3662–3678.
26. Guinee, J.B.; Gorree, M.; Heijungs, R.; Huppes, G.; Kleijn, R.; De Koning, A.; van Oers, L.; Wegener Sleeswijk, A.; Suh, S.; Udo de Haes, H.A.; de Bruijn, H.; van Duin, R.; Huijbregts, M.A.J.; Lindeijer, E.; Roorda, A.A.H.; van der Ven, B.L.; Weidema, B.P. *Handbook on Life Cycle Assessment: Operational Guide to the ISO Standards*; Kluwer Academic Publishers: Leiden, The Netherlands, 2002.
27. McCaskill, M.R.; Ridley, A.M.; Okom, A.; White, R.E.; Andrew, M.H.; Michalk, D.L.; Melland, A.; Johnston, W.H.; Murphy, S.R. SGS Nutrient Theme: environmental assessment of nutrient application to extensive pastures in the high rainfall zone of southern Australia. *Aust. J. Exp. Agric.* 2003, *43*, 927–944.

28. White, R.E.; Christy, B.P.; Ridley, A.M.; Okom, A.E.; Murphy, S.R.; Johnston, W.H.; Michalk, D.L.; Sanford, P.; McCaskill, M.R.; Johnson, I.R.; Garden, D.L.; Hall, D.J.M.; Andrew, M.H. SGS Water Theme: Influence of soil, pasture type and management on water use in grazing systems across the high rainfall zone of southern Australia. *Aust. J. Exp. Agric.* 2003, *43*, 907–926.

29. Sandford, P.; Cullen, B.R.; Dowling, P.M.; Chapman, D.F.; Garden, D.L.; Lodge, G.M.; Andrew, M.H.; Quigley, P.E.; Murphy, S.R.; McG King, W.; Johnson, W.H.; Kemp, D.R. SGS Pasture Theme: Effect of climate, soil factors and management on pasture production and stability across the high rainfall zone of southern Australia. *Aust. J. Exp. Agric.* 2003, *43*, 945–959.

30. Morgan, J.A.; LeCain, D.R.; Pendall, E.; Blumenthal, D.M.; Kimball, B.A.; Carrillo, Y.; Williams, D.G.; Heisler-White, J.; Dijkstra, F.A.; West, M. C_4 grasses prosper as carbon dioxide eliminates desiccation in warmed semi-arid grassland. *Nature* 2011, *476*, 202–205.

31. Eckard, R.J.; Grainger, C.; de Klein, C.A.M. Options for the abatement of methane and nitrous oxide from ruminant production: A review. *Livest. Sci.* 2010, *130*, 47–56.

32. Martin, C.; Morgavi, D.P.; Doreau, M. Methane mitigation in ruminants: From microbe to the farm scale. *Animal* 2010, *4*, 351–365.

33. Cottle, D.J.; Nolan, J.V.; Wiedemann, S.G. Ruminant enteric methane mitigation: A review. *Anim. Prod. Sci.* 2011, *51*, 491–514.

34. Ulyatt, M.J.; Lassey, K.R.; Shelton, I.D.; Walker, C.F. Methane emission from dairy cows and wether sheep fed subtropical grass-dominant pastures in midsummer in New Zealand. *New Zeal. J. Agr. Res.* 2002, *45*, 227–234.

35. Blaxter, K.L.; Clapperton, J.L. Prediction of the amount of methane produced by ruminants. *Brit. J. Nutr.* 1965, *19*, 511–522.

36. Kennedy, P.M.; Charmley, E. Methane yields from Brahman cattle fed tropical grasses and legumes. *Anim. Prod. Sci.* 2012, *52*, 225–239.

37. Alcock, D.J.; Hegarty, R.S. Potential effects of animal management and genetic improvement on enteric methane emissions,

emissions intensity and productivity of sheep enterprises at Cowra, Australia. *Anim. Feed Sci. Technol.* 2011, *166-167*, 749–760.

38. Biswas, W.K.; Graham, J.; Kelly, K.; John, M.B. Global warming contributions from wheat, sheep meat and wool production in Victoria, Australia—A life cycle assessment. *J. Clean. Prod.* 2010, *18*, 1386–1392.
39. Basset-Mens, C.; Ledgard, S.; Boyes, M. Eco-efficiency of intensification scenarios for milk production in New Zealand. *Ecol. Econ.* 2009, *68*, 1615–1625.
40. Peters, G.M.; Rowley, H.V.; Wiedemann, S.; Tucker, R.; Short, M.D.; Schulz, S. Red meat production in Australia: Life cycle assessment and comparison with overseas studies. *Environ. Sci. Technol.* 2010, *44*, 1327–1332.
41. Beauchemin, K.A.; Janzen, H.H.; Little, S.M.; McAllister, T.A.; McGinn, S.M. Mitigation of greenhouse gas emissions from beef production in western Canada—Evaluation using farm-based life cycle assessment. *Anim. Feed Sci. Technol.* 2011, *166-167*, 663–677.
42. Browne, N.A.; Eckard, R.J.; Behrendt, R.; Kingwell, R.S. A comparative analysis of on-farm greenhouse gas emissions from agricultural enterprises in south eastern Australia. *Anim. Feed Sci. Technol.* 2011, *166-167*, 641–652.
43. Eckard, R.J.; Johnson, I.; Chapman, D.F. Modelling nitrous oxide abatement strategies in intensive pasture systems. *Int. Congr. Ser.* 2006, *1293*, 76–85.
44. Dalal, R.; Wang, W.; Robertson, G.; Parton, W. Nitrous oxide emission from Australian agricultural lands and mitigation options: A review. *Aust. J. Soil Res.* 2003, *41*, 165–195.
45. Mitchell, P.J. Gross energy levels in regrowths of six pasture species, and relations with digestibility and chemical composition. *Aust. J. Exp. Agric.* 1974, *14*, 33–37.
46. McDonald, P.; Henderson, A.R.; Ralton, I. Energy changes during ensilage. *J. Sci. Food Agric.* 1973, *24*, 827–834.
47. Donaldson, E.; Edwards, R.A. Feeding value of silage: Silages made from freshly cut grass, wilted grass and formic acid treated wilted grass. *J. Sci. Food Agric.* 1976, *27*, 536–544.

48. Nottle, M.C. Digestible energy of wheat and oat grains when fed to sheep. *Aust. J. Exp. Agric.* 1971, *11*, 610–614. [

49. Intergovernmental Panel on Climate Change, 2006 IPCC Guidelines for National Greenhouse Gas Inventories. In *Agriculture, Forestry and other Land Use*; Eggleston, H.S., Buendia, L., Miwa, K., Ngara, T., Tanabe, K., Eds.; Institute for Global Environmental Strategies (IGES): Hayama, Japan, 2006; Volume 4.

50. Bell, M.; Cullen, B.; Eckard, R. Effect on nitrogen losses from a sheep grazing system by the randomised distribution of excreta. In Proceedings of International Congress on Modelling and Simulation, Perth, Australia, 12–16 December; 2011.

51. Granli, T.; Bøckman, O.C. Nitrous oxide from agriculture. *Norw. J. Agr. Sci. Suppl.* 1994, *12*, 7–128.

52. Vleeshouwer, L.M.; Verhagen, A. Carbon emission and sequestration by agricultural land use: A model study for Europe. *Global Change Biol.* 2002, *8*, 519–530.

53. Wells, C. *Total Energy Indicators of Agricultural Sustainability: Dairy Farming Case Study*; Ministry of Agriculture and Forestry: Wellington, New Zealand, 2001.

54. Lovett, D.K.; Shalloo, L.; Dillon, P.; O'Mara, F.P. A systems approach to quantify greenhouse gas fluxes from pastoral dairy production as affected by management regime. *Agr. Syst.* 2006, *88*, 156–179.

Chapter 2

Carbon Footprints for Food of Animal Origin: What are the Most Preferable Criteria to Measure Animal Yields?

Gerhard Flachowsky[1] and Josef Kamphues[2]

[1]Institute of Animal Nutrition, Friedrich-Loeffler-Institute (FLI), Federal Research Institute for Animal Health, Bundesallee 50, 38116 Braunschweig, Germany

[2]Institute of Animal Nutrition, University of Veterinary Medicine Hannover, Foundation, Bischofsholer Damm 15, 30173 Hannover, Germany

ABSTRACT

There are increasing efforts to determine the origin of greenhouse gas emissions caused by human activities (including food consumption) and to identify, apply and exploit reduction potentials. Low emissions are generally the result of increased efficiency in resource utilization. Considering climate related factors, the emissions of carbon dioxide, methane and laughing gas are summarized to so-called carbon footprints (CF). The CF for food of animal origin such as milk, eggs, meat and fish depend on a number of influencing factors such as animal species, type of production, feeding of animals, animal performance, system boundaries and outputs of production. Milk and egg yields are more clearly defined animal yields or outcomes of production than food from the carcasses of animals. Possible endpoints of growing/slaughter animals are body weight gain, carcass weight gain (warm or cold), meat, edible fractions or edible protein. The production of edible protein of animal origin may be considered as one of the main objectives of animal husbandry in many countries. On the other hand, the efficiency of various lines of production and the CF per product can also be easily compared on the basis of edible protein. The pros and contras of various outputs of animal production under special consideration of edible protein are discussed in the paper.

INTRODUCTION

The current world situation is characterized by a growing population and a higher need for feed and food. These facts are, in turn, associated with a growing demand for limited natural resources such as fuel, land area, water, *etc.*, and with elevated emissions with greenhouse gas (GHG) potential. Such gases can include, e.g., carbon dioxide (CO_2), methane (CH_4), laughing gas (N_2O) and other substances (e.g., N, P, trace elements, *etc.*; (e.g., [1,2,3,4,5]). During the last few years special attention has been devoted to various gases because of their greenhouse gas potential. This increase is discussed in the context of global warming and possible climate change [2].

Agriculture and especially animal husbandry are considered as important greenhouse gas sources because of the high greenhouse potential of their emissions (e.g., $CO_2 \times 1$; $CH_4 \times 23$ and $N_2O \times 296$; [2]). So-called Carbon Footprints (CF; Life cycle assessments (LCA), Eco-Balances) consider the greenhouse gas potential of climate relevant gases and are expressed relative to one gram or kilogram of CO_2 equivalent (CO_{2eq}) per product unit.

Various authors calculated such CF for agriculture in general, but also for separate segments. For example, according to Steinfeld *et al.* [1] livestock production contributes about 18% to the global anthropogenic GHG emissions. O`Mara [6] has reported that animal agriculture is responsible for 8–10.8% of the global GHG emissions. After Lesschen *et al.* [7] livestock farming contributes to global warming with about 10% of total GHG emissions from the EU-27. FAO [4] asserts that the global dairy sector contributes with 3.0 to 5.1% to total anthropogenic greenhouse gases, but Sevenster and Jones [8] calculated only 1.2% from dairy livestock to the global greenhouse gas emissions. The highest variation of global GHG emissions from livestock is mentioned by Herrero *et al.* [9] with a range from 8 to 51%. Detailed information about the GHG emission in the EU is given by Leip *et al.* [10]. From a total of about 660 million tonnes CO_{2eq} per year from livestock, about 65% come from ruminants (production of milk, beef, sheep and goats [10]). Methodical and regional differences make it difficult to compare such values, to make conclusions or to give data based advice to policy makers. The objectives of CF are to sensitize producers and consumers for an efficient use of fossil carbon sources and to reduce greenhouse gas emissions per unit of product.

During the last 10 years many studies dealt with calculations of CF for almost all products resulting from human activities, including production of food of animal origin (e.g., [4,5,7,10,11,12,13,14,15,16,17]). The large range in CF when comparing results of various authors depends on many influencing factors as exemplary shown in Table 1 and Table 2 for milk and beef. The CF for milk varies between 0.4 and 1.5 kg CO_{2eq}/kg milk, taking different world regions into account, between 1.3 (Europe and North America) and 7.5 kg CO_{2eq}/kg in sub-Saharan Africa [4] (see Table 1). Furthermore, most authors considered only the emissions during the production, but sometimes processing, transport and trade are also included in the calculation.

Table 1: Examples of Carbon Footprints (CF) (kg CO_{2eq}/kg milk) depending on the type of production

Type of production/farming			References
Country	Conventional	Organic	
Germany	0.83	0.84	[18]
Germany	0.85	0.78	[14]
Sweden	0.90	0.94	[19]
Germany	0.94	0.88	[13]

The NL	0.97	1.13	[20]
Germany	0.98	0.92	[21]
Sweden	0.99	0.94	[22]
UK	1.06	1.23	[12]
Austria	1.20	1.00	[23]
UK	1.20	1.30	[24]
Germany	1.30	1.30	[25]
The NL	1.40	1.50	[26]
UK	1.6 (1.0–3.2)	1.3 (0.9–2.4)	[27]
Without differentiation in conventional/organic			
Germany	0.40 (40 kg milk/day)		[28]
(model calculation)	0.55 (20 kg milk/day)		[28]
	1.00 (10 kg milk/day)		[28]
Germany	0.65		[29]
New Zealand	0.65–0.75		[30]
Literature review	0.8–1.4 (on farm)		[8]
	0.9–1.8 (on farm + post farm emissions)		
New Zealand	0.86		[31]
Germany	0.98 (10,000)–1.35 (6,000 kg milk/year; see Table 3)		[32]
Sweden	1.00		[33]
Canada	1.00		[34]
UK	1.06		[12]
USA	1.09		[35]
EU-27	1.3 (1.0–2.3)		[7]
Ireland	1.3–1.5		[36]
Global	2.4 (1.3–7.5)		[4]

Still higher variations of CF are described for beef (see Table 2). The values are influenced by body weight gain, feeding, production system and system limits. There are many ways for CF calculation of the yield of growing animals such as body weight gain, hot standard carcass weight, empty body weight, meat, meat plus edible organs or edible protein. In dependence on the calculation basis, the authors found a high variation in CF for beef. The highest values are given for beef cows (Table 2). In general all the results indicate (e.g., [6,37,38]) that activities which are targeted at improvements in productivity and efficiency of resource use will result in a lower GHG emission or lower CF per unit of product.

Table 2: Examples for CF (kg CO_{2eq}/kg carcass weight gain) of beef cattle depending on type of production

Type of production/farming			References
Country	Conventional	Organic	
Germany	8.5	29.0 (beef cow)	[39]
Germany	8.7/10.1	10.2	[18]
Australia	9.9(grain finished)	12.0(grass finished)	[37]
Global	10	32–40	[24]
	(intensive–dairy beef)	(organic–suckler beef)	
Germany	13.3	11.4	[13]
Germany	15.2	17.5	[40]
UK	15.8	18.2	[12]
Ireland	23.6	20.2	[41]
Global	24.5	20.9	[42]
Without differentiation in conventional/organic			
Germany	5.6 (6,000)–14.6 (10,000 kg milk per cow per year, see Table 3)		[32]
Canada	5.9–10.4		[34]
Germany	7.0–23.0		[28]
Germany	8.4 (fattening of calves from dairy cows)		[14]
	16.8 (fattening of calves from beef cows)		
Sweden	10.1		[43]
Ireland	13.0 (11.3–15.6)		[41,44]
Global	15.6 (fattening of calves from dairy cows)		[4]
	20.2 (fattening of calves from beef cows)		
EU	16.9–19.9 (fattening of calves from dairy cows)		[45]
	27.3 (fattening of calves from beef cows)		
Japan	19.6		[46]

Japan	36.4 (beef cows, fattening bulls; 40% meat yield)	[47]

Apart from the factors mentioned above, the allocation of animal products (e.g., [15,26,32,43,48]) may be used whenever systems under study generate more than one saleable output (e.g., milk and meat) or various co-products. Such studies also influence the results of LCA (e.g., [32,37,48]). Mass-based and economic-based allocations were applied. For example Zehetmeier *et al.* [32,38] calculated CF of 1.35 and 0.98 kg CO_{2eq} per kg milk of cows producing 6,000 and 10,000 kg milk per year, respectively. In the case of lower milk yield, beef was produced by calves of dairy cows with a CF of 5.58 kg; in the case of higher milk yields, beef cows are needed to produce sufficient beef and the CF increased to 14.62 kg CO_{2eq} per kg beef. Under consideration of economic aspects (prices for milk and beef; economic allocation), the CF for milk decreased, those of beef increased.

Under consideration of all aspects mentioned above, it is extremely difficult to compare results of LCA from different authors. This variability caused confusion between scientists, among policy makers and in the public. A methodical agreement generated by internationally recognized scientific panels with expertise across a range of disciplines and clear science based orientation (e.g., [9,16,24]) seems to be urgently necessary.

Based on these, the objective of the present review is to characterize the most important influencing factors along the food chain for calculation of CF for food of animal origin. The next section deals with the whole food chain and its system boundaries followed by a critical assessment of the different kinds of animal yields or specific outputs of animal production under special consideration of edible protein and methods of their measurement/assessment.

SETTING THE BOUNDARIES OF A PRODUCTION SYSTEM AND FURTHER INFLUENCING FACTORS ALONG THE FOOD CHAIN

Emissions along the Food Chain

The IPCC [2] recommended GHG factors for CO_2 (1), CH_4 (23) and N_2O (296) to calculate CF for various processes. Recently there was some discussion about the factors and the IPCC [49] revised the global warming potential of methane (CH_4) from 23 to 25 because of indirect effects of CH_4 on ozone and stratospheric water vapour [9]. On the other hand, the Worldwatch Institute [50] suggests a Global Warming Potential for CH_4 of 72. Methane is a very important gas for CF calculation, especially for food derived from ruminants. Between 50 and 80% of the total GHG emissions of food of ruminant origin are due to methane [37,38].

The N_2O-factor is given with around 300 (296 [2]; 298 [49]). From a global view the agricultural N_2O-emissions (from manure and soil; given in billion tonnes of CO_{2eq}/y) are larger than calculated methane emissions (2.5 *versus* 2.15 billion t CO_{2eq}/y, [6]). Analogue tendencies are reported from agriculture in the USA (222 million t CO_{2eq}/y from N_2O; 197 million t CO_{2eq}/y from CH_4; [51]). From the view of science and policy, further research is required to considering a time horizon of the GHG emissions [9,52].

In addition to GHG factors a possible exact measuring of climate relevant gases in all links along the food chain is an essential prerequisite to calculate CF.

There is general agreement that carbon dioxide emissions from livestock metabolism are not considered as a CO_2-source in CF ([2,9], see Figure 1). CO_2 has been fixed by photosynthesis in phytogenic biomass and the C consumed in feed and emitted as CO_2 by animals is considered as equivalent or as emission neutral. On the other hand, CO_2 from technical processes associated with animal husbandry should be considered in CF calculations (for details see [7,10,11]).

Methane can be considered as an unavoidable by-product of anaerobic microbial fermentation, especially in the rumen of ruminants, but also in the hindgut of all species and during anaerobic manure management. Since energy losses via methane in the rumen are well known [53], animal nutritionists have been trying to reduce the gastrointestinal methane emission from ruminants and in the hindgut of further species for a long time. In 2005 around 90 million tonnes CH_4 (about 1.9 billion t CO_2eq/y) were emitted from gastrointestinal fermentation of ruminants [6]. These emissions are projected to grow by over 30% from 2000 to 2020 [6]. Enteric methane emissions (e.g., [54,55,56,57]), methods of measurements (e.g., [58,59,60,61]) and reduction potentials (e.g., [62,63,64,65]) should not be further considered here. Furthermore there are special issues of the journals *"Animal Feed Science and Technology"* 145 (2008), pp. 209–419 and 166–167 (2011), pp. 1–782 and in the *"Australian Journal of Experimental Agriculture"* 48, No. 1 and 2 (2008), dealing with the topics mentioned above.

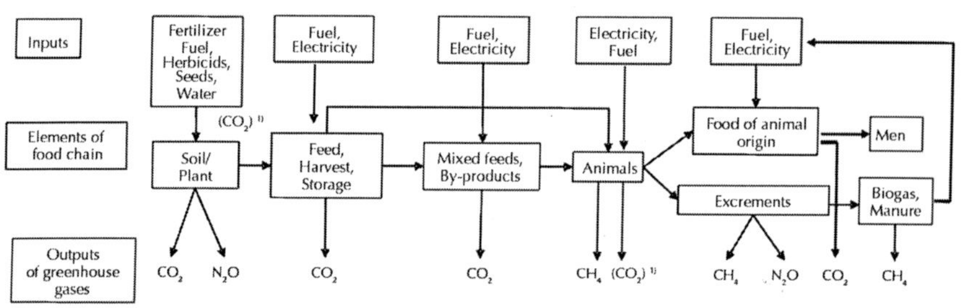

[1)] CO_2 will be fixed by photosynthesis and produced by animal metabolism, therefore it is considered as emaissions-neutral.

Figure 1: Substantial elements of the chain to produce food of animal origin, as well as selected inputs of resources and outputs of greenhouse gases (basic concept for system boundaries [17]).

Laughing gas (N_20) has the highest GHG potential from the most relevant GHG [2]. It is not directly excreted by animals; it depends on microbial conditions during manure management and in the soil. Lower N-excretion by animals [66,67,68] and improved N-management (e.g., storing of manure, adequate amounts ,of manure and fertilizer) may contribute to lower N-emissions from manure and soil (e.g., [69,70,71,72]). More representative results from N_2O measurements (also for grazing animals) may substantially contribute to more reliable data for calculation of CF along the food chain. In some cases specific emissions were excluded from calculations of CF due to data uncertainty (e.g., N_2O from leguminous pastures [37]). Such special situations should be mentioned in the study report to assess the results.

Setting the System Boundaries

Definition of system boundaries along the food chain (see Figure 1 and [24]) is the starting point for GHG measurements, for the calculation of CF of livestock products and for comparing results from various studies [4,10]. There are some open questions which need to be answered and it should be clearly defined whether they are or are not considered in the calculation, such as:

- Consideration of emissions from land use and land use change [1]
- Emissions from basic equipment (e.g., houses, machinery *etc.*; [38,73])
- Transport, processing, trade of products of animal origin [38]
- Emissions during preparing food from animal products in the kitchen/food processing
- Use of an allocation of various animal products (see above)
- For example under Australian conditions Peters *et al.* [37] defined the system boundary in the case of beef for all on-site and upstream processes at the farm, feedlot, and whole processing plant, including transport between these sites.

Table 3 shows exemplary CF for milk under consideration of various boundaries. A clear definition of the system boundaries and the comprehensibility are important prerequisites to follow the calculations and to make the results comparable (e.g., [37,41,44]). Scientists working in this field should come to agreements concerning system boundaries and GHG factors of climate relevant gases.

Table 3: Model calculation to demonstrate the effects of setting different boundaries for CF of milk (g CO_{2eq} per kg milk; 30 kg milk per day; diet on DM-base: 60% roughage, 40% concentrate; 4% milk fat, 3.4% protein; 305 days of lactation; 60 days dry period, 3 years lactation; 30 months calf and heifer period [38]

System	System boundaries	CF (g CO2eq/kg milk)
1	Dairy cow emissions during lactation	280
2	1. + Emissions of feed production	430
3	2. + Dry period	500
4	3. + Heifer period	730
5	4. + Animal housing and milking	760
6	5. + Manure management	820
7	6. + Processing, transportation and trade of milk	1,100

OUTCOME OF ANIMAL PRODUCTION

There is no essential need for food of animal origin for human beings, but the consumption of milk, eggs, meat and fish may substantially contribute to a more balanced and palatable human diet. Therefore one of the main goals of animal husbandry is the production of food of animal origin. Such food contributes substantially to meeting the human requirements in essential amino acids (e.g., [74,75,76]) because of the high content in essential amino acids (such as lysine, methionine and cysteine, threonine, leucine, *etc.*; see [77]). Furthermore, such food contains important minor nutrients like major and trace minerals (such as Ca, P, Cu, Fe, I, Se, Zn) and vitamins (e.g., A, E, some B-vitamins, especially B_{12}) and has a considerable enjoyment value [78]. Human nutritionists [79,80] recommend that about one third of the daily protein requirement (0.66–1 g per kg body weight and day; [75,80,81]) should originate from protein of animal origin to guarantee a more balanced diet, especially for "risk groups" such as pregnant and lactating women, infants and children. Therefore, one endpoint, or the outcome of specific animal yields could be edible protein or essential amino acids and should be clearly defined. Otherwise there will be discrepancies in calculations and variations in the results between various working groups as shown in Table 1 and Table 2.

Milk and Eggs

Milk and eggs are clearly defined as food of animal origin. The yield can be measured as weight (kg, *etc.*) or on the basis of standardized products (e.g., standardized protein, fat, dry matter or energy). Therefore it is relatively easy to measure the animal yield for further calculations. But nevertheless, there is a certain range between CF for milk (see Table 1).

The composition of milk and eggs is well defined (see Table 6), but it may vary between various sources and depending on animal breed, feeding and other influencing factors. Therefore the analysis of milk and egg composition (protein; fat, lactose) may contribute to being more specific in measuring the animal yield incl. the energy yield. Milk and eggs may be used entirely as food (except small amounts of colostrum and egg shells; see Table 7).

Food from Slaughtered Animals

It is much more difficult to quantify and characterize the yield from the animal body after slaughtering and processing. The GHG balance per kilogram body weight gain can only be calculated on the farm level. Mialon *et al.* [82] carried out a feeding study with Blond d'Àquitaine bulls in the finishing phase (400–650 kg body weight) including various feeding systems and weight gains between 1,494 and 1,862 g/d and Doreau *et al.* [83] calculated a CF between 3.6 and 4.7 kg CO_{2eq} per kg body weight gain. Those values are similar to high daily weight gains as shown in Table 5, but much lower than for lower weight gains. Normally, the GHG emissions for the whole beef system include also emissions of cows, calves and heifers, needed to produce beef. They are much higher than in the system dairy cow—growing/fattening bulls for beef (see allocation).

Mostly the term "meat" is used, but it is not clearly described, what it means (real meat or meat plus bones). Peters *et al.* [37] introduced the term "hot standard carcass weight" (HSCW) as the weight at the exit gate of the meat processing plant. It varies between 50–62% of the live weight of the cattle to be slaughtered, but it may vary between 50% in the case of sheep and up to 80% for fattening turkeys (e.g., [12,37,83]).

In the case of animals for meat/fish production the following endpoints can be measured:

- Weight gain of the animal (per day or per growing period) during the whole life span
- Weight gain of animal without gastro-intestinal tract
- Empty body weight (or carcass weight; meat and bones; warm as HSCW or cold)
- Meat (empty body minus bones)
- Edible fraction (meat plus edible organs and tissues)
- Edible protein (edible fractions of the carcass multiplied with their specific protein content).

Therefore it is really difficult to find an adequate CF for meat or edible products from slaughtered animals. Various authors used different bases to calculate CF for products from slaughtered animals. Williams *et al.* [12] estimated the killing out percentages for beef and poultry with 55 and 70% and 72, 75 and 77% for pigs with live weights of 76, 87 and 109 kg, respectively. Lesschen *et al.* [7] used fixed values to calculate the carcass fraction from the final body weight of animals (e.g., 58% for beef; 75% for pork and 71% for poultry). Most authors used a fixed fraction of 0.9 for all animal species for conversion of carcass weight to edible "meat". De Vries and de Boer [5] used calculation factors to determine the amount of edible product per kg live weight by 0.43; 0.53 and 0.56 for beef, pork and poultry. Table 4 shows potential outputs for growing/fattening cattle under consideration of various endpoints as mentioned above.

Calculation of CF may base on various outputs. For practical reasons carcass weight or weight gain (warm or cold) would be the most important endpoint to measure the yield of slaughtered animals because this weight is measurable in the abattoir [37] and can be used for further calculations. Based on the values derived from Table 4, CF is calculated for various endpoints under consideration of differences in feeding and greenhouse gas emissions and is shown in Table 5.

Table 4: Model calculation to show various endpoints for growing/fattening bulls (150-550 kg body weight; calculation based on data collected by [84]

Gross weight gain (g/day)	Weight gain without content of intestinal tract (g/day)	Carcass weight (warm; % of weight gain)	Carcass weight gain (warm; g/day)	Meat gain (% of weight gain)	Meat gain (g/day)	Edible fraction gain 1 (g/day)	Edible protein (g/day; 19% protein in edible fraction)
500	438	50	250	40	200	250	48
1,000	900	53	530	44	440	490	93
1,500	1,385	56	840	48	720	770	146

[1]Meat plus other edible tissues

Table 5: Model calculations for CF of beef (150-550 kg body weight [1]) depending on feeding, weight gain, methane- and N_2O-emissions and N-excretion [28]

Weight gain (g/day)	Feed intake (kg DM/ (animal x day)	Portion concentrate (% of DM-intake) 1,2	Methane emissions (g/kg DM)	N-excretion (g/day)	N2O-synthesis (% of N-excretion)	Carbon footprints (kg CO2eq/kg)			
						Weight gain	Empty carcass weight gain	Edible fraction gain	Edible Protein
500 (Pasture, no concentrate)	6.5	0	26	110	2	11.5	23.0	28.0	110
1,000 (Indoor, grass silage, some concentrate)	7.0	15	24	130	1	5.5	11.0	13.8	55
1,500 (Indoor, corn silage, concentrate)	7.5	30	22	150	0.5	3.5	7.0	9.0	35

[1]Production of calf up to 150 kg BW is not considered; 2 CO2-Emission: 120 g/kg roughage-DM; 220 g/kg concentrate-DM.

Edible Protein as Most Important Objective of Animal Husbandry

The production of protein of animal origin is one of the most important goals of animal husbandry [5]. On the other hand, the efficiency and the emissions of animal products can be also compared on the basis of edible protein. The N or protein content of various foods of animal origin may vary from values used for calculations in Table 6 (data by [84] on the basis of own studies). Our data agrees with values used by Lesschen *et al.* [7], and it does not substantially disagree with values from human food tables (see Table 6). De Vries and de Boer [5] used for their calculations 190 g protein/kg edible beef, pork and poultry meat; 30 g per kg milk products and 130 g per kg eggs.

Considering various influencing factors such as animal yields, feeding, edible fractions and protein content in the edible fractions, the yield of edible protein per day and per kg body weight of animals is given in Table 7.

Table 6: Published data regarding the protein content of some edible animal products (in g per kg edible product)

Product	References				
	[7]1	[77]	[84]	[85]	[86,87,88,89]
Cows milk	34.4	33.3 (30.8-37.0)	32	34	34
Beef	206	220 2 (206-227)	190	206-212	170-200
Pork	156	220 2 (195-240)	150	183-216	157 (129-178)
Broiler	206	199	200	182-242	n.d.
Eggs	119	125	120	125	121 (110-124)

[1]N-content × 6.25; [2]Muscles only; n.d.: no data

The feeding may influence CF of food of animal origin. In the case of ruminants, higher amounts of concentrate are required for higher animal yields. The proportion of by-products [90,91] used in animal feeding does not only have nutritional implications, but it also affects the results of calculations on land use [92]. There are large differences in protein yield per animal per day or per kg body weight and day depending on animal species and category as well as their performances and the fractions considered as edible (see Table 7).

Table 7: Influence of animal species, categories and performances on yield of edible protein [84]

Protein source (Body weight)	Performance per day	Dry matter intake (kg per day)	Roughage to concentrate ratio (on DM base, %)	Edible fraction (% of product or body mass)	Protein in edible fraction (g per kg fresh matter)	Edible protein (g per day)	Edible protein (g per kg body weight and day)
Dairy cow (650 kg)	10 kg milk	12	90/10	95	34	323	0.5
	20 kg milk	16	75/25			646	1.0
	40 kg milk	25	50/50			1292	2.0
Dairy goat (60 kg)	2 kg milk	2	80/20	95	36	68	1.1
	5 kg milk	2.5	50/50			170	2.8
Beef cattle (350 kg)	500 g 1	6.5	95/5	50	190	48	0.14
	1,000 g 1	7.0	85/15			95	0.27
	1,500 g 1	7.5	70/30			143	0.41
Growing/ fattening pig (80 kg)	500 g 1	1.8	20/80	60	150	45	0.56
	700 g 1	2	10/90			63	0.8
	1,000 g 1	2.2	0/100			81	1.0
Broiler (1.5 kg)	40 g 1	0.07	10/90	60	200	4.8	3.2
	60 g 1	0.08	0/100			7.2	4.8
Laying hen (1.8 kg)	50% 2	0.10	20/80	95	120	3.4	1.9
	70% 2	0.11	10/90			4.8	2.7
	90% 2	0.12	0/100			6.2	3.4

1Daily weight gain, 2Laying performance.

Table 7 shows the highest protein yields per kg body weight for growing broilers as well as for laying and lactating animals and the lowest values for growing/fattening ruminants. Based on those values, emissions per kg edible protein are given in Table 8. Higher proportions of edible fractions or higher protein content (e.g., 50 g protein per kg camel milk, [77]) as shown in Table 6 and Table 7 may increase the protein yield and reduce the CF per unit of product.

Apart from protein food of animal origin also contains fat and some carbohydrates which contribute to human nutrition and which may replace energy of plant origin in human diets.

At high levels of performance there are remarkable differences in CO_2 emissions due to human consumption of 1 g protein from food of animal origin (eggs and meat from broiler < pork < milk < beef, see Table 7). But here it has to be emphasized that this protein intake is accompanied—willingly or not—by an energy intake from the protein itself, but also from further nutrients like lactose and fat in milk or from fat in meat or eggs. Therefore, it should be avoided to attribute the CO_2 burden to the protein fraction ("edible protein") exclusively. To prevent that this fact is neglected, there are different alternatives:

In a first simple method, the CO_2 emission due to 1 kg edible protein could be used as CO_2 burden of consumed energy (for example: 1 kg edible protein of eggs corresponds to about 8 kg egg corresponding to 51.6 MJ energy (calculated by [77]); these combined intakes are related to 3 kg CO_2).

One alternative could be a "nutritional allocation" (as described before for economic allocation), meaning that the CO_2 emissions are attributed to different functions of the food (source of protein/source of energy/source of further essential nutrients).

Table 8: Influence of animal species, categories and performances on emissions (per kg edible protein, own calculations)

Protein source (Body weight)	Performance per animal per day	N-excretion (% of intake)	Methane emission (g per day)[3]	Emissions in kg per kg protein			
				P	N	CH4[3]	CO2eq
Dairy cow (650 kg)	10 kg milk	75	310	0.10	0.65	1.0	30
	20 kg milk	70	380	0.06	0.44	0.6	16
	40 kg milk	65	520	0.04	0.24	0.4	12
Dairy goat (60 kg)	2 kg milk	75	50	0.08	0.5	0.8	20
	5 kg milk	65	60	0.04	0.2	0.4	10
Beef cattle (350 kg)	500 g[1]	90	170	0.30	2.3	3.5	110
	1,000 g[1]	84	175	0.18	1.3	1.7	55
	1,500 g[1]	80	180	0.14	1.0	1.2	35

Growing/ fattening pig (80 kg)	500 g 1	85	5	0.20	1.0	0.12	16
	700 g 1	80	5	0.12	0.7	0.08	12
	900 g 1	75	5	0.09	0.55	0.05	10
Broilers (1.5 kg)	40 g 1	70	Traces	0.04	0.35	0.01	4
	60 g 1	60		0.03	0.25	0.01	3
Laying hen (1.8 kg)	50% 2	80	Traces	0.12	0.6	0.03	7
	70% 2	65		0.07	0.4	0.02	5
	90% 2	55		0.05	0.3	0.02	3

[1]Daily weight gain [2]Laying performance [3]CH4-emission varies with composition of diet

In a first simple step it is recommended to diminish the CO_2 emission per 1 kg edible protein (Table 8) by the CO_2 amounts that would occur at an identical energy intake from food of plants (energy from carbohydrates and fat). It means that an intake of 1 kg protein from eggs (corresponds to 8 kg eggs; see Table 7; and corresponding to 51.6 MJ energy) saves high amounts of other food (and their CO_2 burden). A more sophisticated way of an "allocation" within the foods could be to differentiate between "protein derived energy" and "non-protein derived energy". In milk and eggs more than 50% of the total energy content is related to the non-protein-fraction (lactose/fat), therefore, it is questionable whether the entire CO_2 emission should be attributed only to the protein intake. Due to the very low CO_2 emission caused by energy intake of carbohydrates and fat from plants/seeds [12,15,17] it would avoid/save high amounts of CO_2 emissions, if the production of food of animal origin focussed on "edible protein" and not on energy of non-proteinaceous fractions.

Furthermore animal products are not only used as food or respectively, as protein/amino acids, and energy sources; they also offer some other important side-products such as skins or hides, fish meal or meat and bone meal, *etc.* A kind of combined "nutritional/further purposes allocation" may contribute to a more scientific assessment of CF for nutrient and energy supply as well as further uses.

Advantages and weaknesses of endpoints (outputs) of various types of animal production are summarized in Table 9. All endpoints are characterized by some advantages and disadvantages. From nutritional and scientific points of view the edible protein seems to be the most favourable measurement, but its measurement is not easy and requires some analytical work (see Table 9). Land requirements (e.g., arable land

and/or grassland) as well types and intensities of food production may be calculated on the basis of various protein sources for human nutrition. Such calculations can contribute for better understanding of various conflicting aims in the field of food production, human nutrition, use of unlimited and limited resources and resource efficiency, emissions and further points in public discussion.

Table 9: Advantages and disadvantages of various outputs/endpoints of animal yields

Animal yields	Advantages	Disadvantages
Milk, Eggs	Easily measurable, almost complete edible	Variation in protein, fat and energy yield, analyses may be useful
Body weight gain	Easily measurable	High portion of non edible fractions in the gains
Carcass weight	Easily measurable	Contains still fractions which are not edible (e.g., bones)
Meat, edible fraction	Completely edible	Categorization and separation not easy
Edible protein	Most important objective of animal production; comparison of various methods and sources to produce protein of animal origin	Categorization of various fractions as edible and difficulties to measure; additional analytical work; variation in N/protein content

CONCLUSIONS

Ranking of food of animal origin on the basis of CF may be indicative for some products, but may also lead to wrong conclusions because of incompleteness of measuring animal yields and data bases, system boundaries and other weaknesses. The data bases for GHG emissions should be improved and the animal yields should be made comparable. Edible protein (or rather, essential amino acids and some minor nutrients) of animal origin, being the most important objective of animal husbandry, is proposed as a standard to compare various types and intensities of animal husbandry. Furthermore we have to look at the whole food chain (see Figure 1) in order to decide whether a practice is sustainable or not in the long term (e.g., [93,94]). In order to do this, however, further research is needed for more reliable and resilient data.

ACKNOWLEDGMENTS

The authors would like to thank to Manfred Grün for critical comments; Anne Mösseler for some calculations; and Saara Sander and Sigrid Herweg for technical assistance.

REFERENCES

1. Steinfeld, D.; Gerber, P.; Wassenaar, T.; Castel, V.; Rosales, M.; de Haan, C. *Livestock's Long Shadow. Environmental Issues and Options*; FAO: Rome, Italy, 2006.
2. *IPCC Guidelines for National Greenhouse Gas Inventories. Agriculture, Forestry and Other Land Use*; Intergovernmental Panel on Climate Change: Cambridge, UK, 2006; Volume 4.
3. *The State of Food and Agriculture—Livestock in the Balance*; FAO: Rome, Italy, 2009.
4. *Greenhouse Gas Emissions from the Dairy Sector. A Life Cycle Assessment*; FAO: Rome, Italy, 2010.
5. De Vries, M.; de Boer, I.J.M. Comparing environmental impacts for livestock products: A review of life cycle assessments. *Livest. Sci.* 2010, *128*, 1–11.
6. O'Mara, F.P. The significance of livestock as a contributor to global greenhouse gas emissions today and in near future. *Anim. Feed Sci. Technol.* 2011, *166–167*, 7–15.
7. Lesschen, J.P.; van den Berg, M.; Westhoek, H.J.; Witzke, H.P.; Oenema, O. Greenhouse gas emission profiles of European livestock sectors. *Anim. Feed Sci. Technol.* 2011, *166–167*, 16–28.

8. Sevenster, M.; de Jong, F. *A Sustainable Dairy Sector—Global, Regional and Live Cycle Facts and Figures on Greenhouse-Gas Emissions; Technical Report; Publication No.: 08.7789*; CE Delft: Delft, The Netherlands, 2008.
9. Herrero, M.; Gerber, P.; Vellinga, T.; Garnett, T.; Leip, A.; Opio, C.; Westhoek, H.J.; Thornton, P.K.; Olesen, J.; Hutchings, N.; *et al.* Livestock and greenhouse gas emissions: The importance of getting the numbers right. *Anim. Feed Sci. Technol.* 2011, *166–167*, 779–782.
10. Leip, A.; Weiss, F.; Wasenaar, T.; Perez, I.; Fellmann, T.; Loudjami, P.; Tuiello, F.; Grandgirard, D.; Monni, S.; Biala, K. *Evaluation of the Livestock Sector's Contribution to the EU Greenhouse Gas Emissions (GGELS); Final Report*; European Commission, Joint Research Centre: Ispra, Italy, 2010.
11. Bockisch, F.J.; Ahlgrimm, H.J.; Böhme, H.; Bramm, A.; Dämmgen, U.; Flachowsky, G.; Heinemeyer, O.; Höppner, F.; Murphy, D.P.L.; Rogasik, J.; *et al.* Assessment of types of organic and conventional farming under consideration of energy use and emissions (in German). *Landbauf. Völkenrode* 2000, *SH 211*, 206.
12. Williams, A.G.; Audsley, E.; Sanders, D.L. *Determining the Environmental Burdens and Resource Use in the Production of Agricultural and Horticultural Commodities; Main Report; Defra Research Project IS0205*; Defra, Cranfield University: Bedford, UK, 2006.
13. Fritsche, R.; Eberle, U. *Greenhouse Gas Emissions During Production and Processing of Food (in German); Working Document*; Öko-Institut e.V.: Darmstadt, Germany, 2007.
14. Hirschfeld, J.; Weiß, J.; Precht, M.; Korbun, T. *Climate Effects of Agriculture in Germany (in German)*; Schriftenreihe des IÖW 186/08: Berlin, Germnay, 2008.
15. Grünberg, J.; Nieberg, H.; Schmidt, T. Carbon Footprints of Food; A Critical Reflection (in German). *Landbauforsch. vTI Agric. For. Res.* 2010, *60*, 53–72.
16. IDF (International Dairy Federation). A common carbon footprint approach for dairy—The IDF guide to standard lifecycle assessment methodology for the dairy sector. *Bull. Int. Dairy Fed.* 2010, *445*. IDF No. 445/2010.

17. Flachowsky, G. Carbon-footprints for food of animal origin, reduction potentials and research need. *J. Appl. Anim. Res.* 2011, *39*, 2–14.
18. Woitowicz, A. Consequences of Reduction of Consumption of Food of Animal Origin on Selected Indicators of Sustainability—Demonstrated by Conventional and Organic Farming (in German). Ph.D. Thesis, TU München, München, Germany, 2007.
19. Cederberg, C.; Flysjö, A. *Life Cycle Inventory of 23 Dairy Farms in South-Western Sweden; SIK Rapport 728*; SIK: Goeteborg, Sweden, 2004.
20. Van der Zijpp, I.A.J. Animal Production Systems: On Integration and Diversity. Habilitation Thesis, University of Wageningen, Wageningen, The Netherlands, 2001.
21. GEMIS. *Global Emissions-Model of Integrated Systems (in German); Version 4.7*, 2010. Available online: http://www.oeko.de/service/gemis (accessed on 14 February 2012).
22. Cederberg, C.; Mattsson, B. Life cycle assessment of milk production—A comparison of conventional and organic farming. *J. Clean. Prod.* 2000, *8*, 250–260.
23. Lindenthal, T.; Markut, T.; Hörtenhuber, S.; Rudolph, G.; Hanz, K. Climate advantages once more demonstrated: Climate balance of organic products (in German). *Ökologischer Landbau* 2010, *153*, 51–53.
24. PAS (Publicly Available Specification) 2050, *Scenario Building to Test and Inform the Development of a BSI Method for Assessing Greenhouse Gas Emissions from Food; ADAS Report; Project No: YAW3408*; Defra, Cranfield University: Bedford, UK, 2011.
25. Haas, G.; Wetterich, F.; Köpke, U. Comparing intensive, extensified and organic grassland farming in Southern Germany by process life cycle assessment. *Agric. Ecosyst. Environ.* 2001, *83*, 43–53.
26. Thomassen, M.A.; van Calker, K.J.; Smits, M.C.J.; Iepema, G.L.; de Boer, I.J.M. Life cycle assessment of conventional and organic milk production in the Netherlands. *Agric. Syst.* 2007, *93*, 95–107.
27. Plassmann, K.; Edwards-Jones, G. *Where does the Carbon Footprint Fall? Development a Carbon Map of Food Production;*

Sustainable Markets Discussion; International Institute for Environment and Development: London, UK, 2009.

28. Flachowsky, G. Greenhouse gases and resource efficiency. Aspects of production of food of animal origin (in German). *Ernährungsumschau* 2008, *55*, 414–419.
29. Daemmgen, U.; Haenel, H.D. Emissions of greenhouse gases and gaseous air pollutants—A challenge for animal nutrition. *Proc. Soc. Nutr. Physiol.* 2008, *17*, 163–167.
30. Basset-Mens, C.; Ledgard, S.; Boyes, M. Eco-efficiency of intensification scenarios for milk production in New Zealand. *Ecol. Econ.* 2009, *68*, 1615–1625.
31. Ledgard, S.; Finalyson, J.; Patterson, M.; Carran, R.; Wedderburn, M. Effects of Intensification of Dairy Farming System in New Zealand on Whole-System Resource Use Efficiency and Environmental Emissions. In In Life Cycle Assessment in the Agri-Food Sector; DIAS Report; In Proceedings from the 4th International Conference, Bygholm, Denmark, 6–8 October 2003; pp. 226–229.
32. Zehetmeier, M.; Bandracco, J.; Hoffmann, H.; Heißenhuber, A. Does increasing milk yield per cow reduce greenhouse gas emissions? A system approach. *Animal* 2011, *6*, 154–166.
33. Cederberg, C.; Sonesson, U.; Henriksson, M.; Sund, V.; Davis, J. *Greenhouse Gas Emissions from Swedish Production of Meat, Milk and Eggs 1990 and 2005*; The Swedish Institute for Food and Biotechnology: Gothenburg, Sweden, 2009.
34. Verge, X.P.C.; Dyer, J.A.; Desjardidins, R.L.; Worth, D. Greenhouse-gas emissions from the Canadian dairy industry in 2001. *Agric. Syst.* 2007, *94*, 683–693.
35. Phetteplace, H.; Johnson, D.; Seidl, A. Greenhouse-gas emissions from simulated beef and dairy livestock systems in the United States. *Nutr. Cycl. Agroecosystems* 2001, *60*, 99–102.
36. Casey, J.W.; Holden, N.M. Analysis of greenhouse gas emissions from the average Irish milk production system. *Agric. Syst.* 2005, *86*, 97–114.
37. Peters, G.M.; Rowley, H.S.; Wiedemann, S.; Tucker, R.; Short, M.D.; Schulz, M. Red meat production in Australia: Life cycle assessment and comparisons with overseas studies. *Environ. Sci. Technol.* 2010, *44*, 1327–1332.

38. Flachowsky, G.; Brade, W.; Feil, A.; Kamphues, J.; Meyer, U.; Zehetmeier, M. Carbon (CO_2)-footprints of producing food of animal origin—Data basis and reduction potentials (in German). *Übers. Tierernährg.* 2011, *39*, 1–45.

39. Reitmayr, T. Development of a computerized data system for economical and ecological assessment of agricultural systems (in German). *Agrarwirtsch. Frankfurt/Main* 1995, SH 147.

40. Schlich, E.H.; Fleissner, U. The ecology of scale: Assessment of regional energy turnover and comparison with global food. *Int. J. Life Cycle Assess.* 2005, *10*, 219–223.

41. Casey, J.W.; Holden, N.M. Greenhouse gas emission from conventional, agri-environmental scheme and organic Irish Suckler—Beef units. *J. Environ.Qual.* 2006, *35*, 231–239.

42. Subak, S. Global environmental cost of beef production. *Environ. Econ.* 1999, *30*, 79–91. [Google Scholar]

43. Cederberg, C.; Stadig, M. System expansion and allocation in life cycle assessment of milk and beef production. *Int. J. Life Cycle Assess.* 2003, *8*, 360–356.

44. Casey, J.W.; Holden, N.M. Quantification of GHG emissions from suckler-beef production in Ireland Agric. *Syst.* 2006a, *90*, 79–98.

45. Nguyen, T.L.T.; Hermansen, J.E.; Mogensen, L. Environmental consequences of different beef production systems in the EU. *J. Cleaner Prod.* 2010, *18*, 756–766.

46. Ogino, A.; Kaku, K.; Osada, T.; Shimada, K. Environmental of the Japanese beef-fattening system with different lengths as evaluated by a life-cycle assessment method. *J. Anim. Sci.* 2004, *72*, 2115–2122.

47. Ogino, A.; Orito, H.; Shimada, K.; Hirooka, H. Evaluating environmental impacts of the Japanese beef cow-calf system by the life cycle assessment method. *J. Anim. Sci.* 2007, *78*, 424–432.

48. Flysjö, A.; Cederberg, C.; Henriksson, M.; Ledgard, S. How does co-product handling affect the carbon footprint of milk? Case study of milk production in New Zealand and Sweden. *Int. J. Life Cycle Assess.* 2011, *16*, 420–430.

49. *Climate Change 2007: Synthesis Report; Contribution of Working Groups I, II and III to the Forth Assessment Report of the Intergovernmental Panel on Climate Change, Core Writing Team;* Pachauri, R.K., Reisinger, A., Eds.; IPCC: Geneva, Switzerland, 2007.
50. Goodland, R.; Anhang, J. *Livestock and Climate Change. What if the Key Actors in Climate Change Were Pigs, Chickens and Cows?; Worldwatch Nov./Dec. 2009*; Worldwatch Institute: Washington, DC, USA, 2009; pp. 10–19. U.S. *Greenhouse Gas Inventory Report: 1990–2009*; US Environment Protection Agency: Washington, DC, USA, 2011. Available online: http://www.epa.gov/climatechange/emissions/usinventoryreport.html/ (accessed on 5 August 2011).
51. Plattner, G.K.; Stocker, T.; Midgley, P.; Tignor, M. *IPCC Expert Meeting on the Science of Alternative Metrics: Meeting Report*; IPCC Working Group: Bern, Switzerland, 2009.
52. Tappeiner, H. *The Nutrition of Domestic Animals (in German)*; 7th ed.; Verlagsbuchhandlung Paul Parey: Berlin, Germany, 1916; p. 93. Kebreab, E.; Johnson, K.A.; Archibeque, S.L.; Pape, D.; Wirth, T. Model for estimating enteric methane emissions from United States dairy and feedlot cattle. *J. Anim. Sci.* 2008, *86*, 2738–2748.
53. Janssen, P.H. Influence of hydrogen on rumen methane formation and fermentation balances through microbial growth kinetics and fermentation thermodynamics. *Anim. Feed Sci. Technol.* 2010, *160*, 1–22.
54. Martin, C.; Morgavi, D.P.; Doreau, M. Methane mitigation in ruminants: From microbe to the farm scale. *Animal* 2010, *4*, 351–365.
55. Bannink, A.; van Schijdel, M.W.; Dijkstra, J. A model of enteric fermentation in dairy cows to estimate methane emissions for the Dutch National Inventory Report using the IPCC Tier 3 approach. *Anim. Feed Sci. Technol.* 2011, *166–167*, 603–618.
56. Johnson, K.A.; Huyker, M.; Westberg, H.; Lamb, B.; Zimmerman, P. Measurement of methane emissions from ruminant livestock using a SF6 tracer technique. *Environ. Sci. Technol.* 1994, *28*, 359–362.

57. Makkar, H.P.S.; Vercoe, P.E. *Measuring Methane Production from Ruminants*; Springer: Dordrecht, The Netherlands, 2007.

58. Pinares-Patino, C.S.; Clark, H. Reliability of the sulphur hexafluoride tracer technique for methane emission measurements from individual animals: An overview. *Aust.J. Exp. Agric.* 2008, *48*, 223–229.

59. Place, S.E.; Pan, Y.; Zhao, Y.; Mitloehner, F.M. Construction and operation of a ventilated hood system for measuring greenhouse gas and volatile organic compound emissions from cattle. *Animals* 2011, *1*, 433–446.

60. Flachowsky, G.; Brade, W. Potentials of reduction of methane emissions in ruminants (in German). *Züchtungskunde* 2007, *79*, 417–465.

61. Tamminga, S.; Bannink, A.; Dijkstra, J.; Zom, R. *Feeding Strategies to Reduce Methane Losses in Cattle*; Animal Science Group, Wageningen UR: Wageningen, The Netherlands, 2007.

62. Place, S.E.; Mitloehner, F.M. Invited review: Contemporary environments issues: A review of the dairy industry's role in the climate change and air quality and the potential of mitigation through improved production efficiency. *J. Dairy Sci.* 2010, *93*, 3407–3416.

63. Cooprider, K.L.; Mitloehner, F.M.; Famula, T.R.; Kebreab, E.; Zhao, Y.; van Eenennaam, A.J. Feedlot efficiency implications on greenhouse gas emissions and sustainability. *J. Anim. Sci.* 2011, *89*, 2643–2656.

64. Flachowsky, G.; Lebzien, P. Possibilities for reduction of Nitrogen (N) excretion from ruminants and the need for further research—A review. *Landbauf. Völkenrode* 2006, *56*, 19–30.

65. Arriaga, H.; Salcedo, G.; Martinez-Suller, S.; Calsamiglia, S.; Merino, P. Effect of dietary crude protein modification on ammonia and nitrous oxide concentration on a tie-stall dairy barn floor. *J. Dairy Sci.* 2010, *93*, 3158–3165.

66. Calsamiglia, S.; Ferret, A.; Reynolds, C.K.; Kristensen, N.B.; van Vuuren, A.M. Strategies for optimizing nitrogen use by ruminants. *Animal* 2010, *4*, 1184–1196.

67. Schils, R.L.M.; Verhagen, A.; Aarts, H.F.M.; Kuikman, P.J.; Sebek, L.B.J. Effect of improved nitrogen management on greenhouse

gas emissions from intensive dairy systems in the Netherlands. *Glob. Change Biol.* 2006, *12*, 382–391.

68. Flechard, C.R.; Ambus, P.; Skiba, U.; Rees, R.M.; Hensen, A.; van Amstel, A.; Pol-van Dasselaar, A.V.; Soussana, J.F.; Jones, M.; Clifton-Brown, J.; *et al.* Effects of climate and management intensity on nitrous oxide emissions in grassland systems across Europe. *Agric. Ecosyst. Environ.* 2007, *121*, 135–152.
69. Reidy, B.; Webb, J.; Misselbrook, T.H.; Menzi, H.; Luesink, H.H.; Hutchings, N.J.; Eurich-Menden, B.; Döhler, H.; Daemmgen, U. Comparison of models used for national agricultural emission inventories in Europe: Litter based manure systems. *Atmos. Environ.* 2009, *43*, 1632–1640.
70. Daemmgen, U.; Haenel, H.-D.; Roesemann, C.; Brade, W.; Kueller-Lindenlauf, M.; Eurich-Menden, B.; Döhler, H.; Hutchings, N.J. An improved data base for the description of dairy cows in the German agricultural emission model GAS-EM. *Landbauforsc. vTI Agric. For. Res.* 2010, *60*, 87–100.
71. Brunsch, R.; Kraatz, S.; Berg, W.; Rus, C. Calculation of energy efficiency in animal husbandry on the base of energy balances (in German). *KTBL Schrift* 2008, *463*, 115–125.
72. Young, V.R.; Bier, D.M.; Pellett, P.L. A theoretical basis for increasing current estimates of the amino acid requirements in adult man, with experimental support. *Am. J. Clin. Nutr.* 1989, *50*, 80–92.
73. WHO, FAO and UNU. *Protein and Amino Acid Requirements in Human Nutrition; Report of a Joint WHO/FAO/UNU Expert Consultation; World Health Organization Technical Report Series*; WHO: Geneva, Switzerland, 2007; Volume 935, pp. 1–265.
74. Pillai, R.R.; Kurpad, A.V. Amino Acid Requirements: Quantitative Estimates. In *Amino Acids in Human Nutrition and Health*; D'Mello, J.P.F., Ed.; CAB International: Wallingford, CT, USA, and Cambridge, MA, USA, 2011; pp. 267–290.
75. Souci, S.W.; Fachmann, W.; Kraut, H. *Food Composition and Nutrition Tables*, 7th ed.; Medpharm: Alexandria, VA, USA, 2008.
76. Keyzer, M.A.; Merbis, M.D.; Pavel, I.F.P.W.; van Wesenbeeck, C.F.A. Diet shifts towards meat and the effects on cereal use: Can we feed the animals in 2030? *Ecol. Econ.* 2005, *55*, 187–202.

77. Waterlow, J.R. The mysteries of nitrogen balance. *Nutr. Res. Rev.* 1999, *12*, 25–54
78. Jackson, A.A. Protein. In *Essentials in Human Nutrition*, 3rd ed.; Mann, J., Truswell, S., Eds.; Oxford University Press: Oxford, UK, 2007; pp. 53–72.
79. Rand, W.M.; Pellett, P.L.; Young, V.R. Meta-analysis of nitrogen balance studies for estimating protein requirements in healthy adults. *Am. J. Clin. Nutr.* 2003, *77*, 109–127.
80. Mialon, M.M.; Martin, C.; Garcia, F.; Menassol, J.B.; Dubroeucq, H.; Veissier, I.; Micol, D. Effects of the forage-to-concentrate ratio in the diet on feeding behaviour in young Blond d'Aquitaine bulls. *Animal* 2008, *2*, 1682–1691.
81. Doreau, M.; van der Werf, H.M.G.; Micol, D.; Dubroeucq, H.; Agabriel, J.; Rochette, Y.; Martin, C. Enteric methane production and greenhouse gases balance of diets differing in concentrate in the fattening phase of a beef cattle production system. *J. Anim. Sci.* 2011, *89*, 2518–2528.
82. Flachowsky, G. Efficiency of energy and nutrient use in the production of edible protein of animal origin. *J. Appl. Anim. Res.* 2002, *22*, 1–24.
83. Souci, S.W.; Fachmann, W.; Kraut, H. Food Table for the Practise. In *The Little Souci-Fachmann-Kraut (in German)*, 5th ed.; Deutsche Forschungsanstalt für Lebensmittelchemie, Freising, Wissenschaftliche Verlagsgesellschaft mbH: Stuttgart, Germany, 2011.
84. GfE (Gesellschaft für Ernährungsphysiologie), *Energy and Nutrient Requirements of Domestic Animals. No. 6. Recommendations for Energy and Nutrient Requirements of Beef Cattle (in German)*; DLG-Verlags-GmbH: Frankfurt, Germany, 1995.
85. GfE (Gesellschaft für Ernährungsphysiologie), *Recommendations for Energy and Nutrient Requirements of Laying Hens and Broilers (in German)*; DLG-Verlags GmbH: Frankfurt, Germany, 1999; Volume 7.
86. GfE (Gesellschaft für Ernährungsphysiologie), *Recommendations for Energy and Nutrient Requirements of Dairy Cattle and Heifers (in German)*; DLG-Verlags GmbH: Frankfurt, Germany, 2001; Volume 8.

87. GfE (Gesellschaft für Ernährungsphysiologie), *Recommendations for Energy and Nutrient Requirements of Pigs (in German)*; DLG-Verlags GmbH: Frankfurt, Germany, 2006; Volume 10.
88. Flachowsky, G.; Kamphues, J. Non-conventional feedstuffs (in German). In Proceedings of the Workshop on Unkonventionelle Futtermittel, Braunschweig, Germany, 10–11 April 1996.
89. Kamphues, J.; Flachowsky, G. Animal nutrition—Resources and new challenges (in German). In Procceedings of the Workshop on Tierernährung—Ressourcen and neue Aufgaben, Braunschweig, Germany, 15–16 June 2000.
90. Vandehaar, M.J. Efficiency of nutrient use and relationship to profitability on dairy farms. *J. Dairy Sci.* 1998, *81*, 272–282.
91. Flachowsky, G.; Hachenberg, S. CO_2-footprints for food of animal origin—Present stage and open questions. *J. Cons. Protect. Food Saf.* 2009, *4*, 190–198.
92. Janzen, H.H. What place for livestock on a re-greening earth? *Anim. Feed Sci. Technol.* 2011, *166–167*, 783–796.

Chapter 3

Livestock Helminths in a Changing Climate: Approaches and Restrictions to Meaningful Predictions

Naomi J. Fox[1,2,3], Glenn Marion[3], Ross S. Davidson[1], Piran C. L. White[2] and Michael R. Hutchings[1]

[1]SAC, West Mains Road, Edinburgh, EH9 3JG, UK;

[2]Environment Department, University of York, Heslington, York, YO10 5DD, UK;

[3]Biomathematics and Statistics Scotland, Kings Buildings, Edinburgh, EH9 3JZ, UK

ABSTRACT

Climate change is a driving force for livestock parasite risk. This is especially true for helminths including the nematodes *Haemonchus contortus, Teladorsagia circumcincta, Nematodirus battus*, and the trematode *Fasciola hepatica,* since survival and development of free-living stages is chiefly affected by temperature and moisture. The paucity of long term predictions of helminth risk under climate change has driven us to explore optimal modelling approaches and identify current bottlenecks to generating meaningful predictions. We classify approaches as correlative or mechanistic, exploring their strengths and limitations. Climate is one aspect of a complex system and, at the farm level, husbandry has a dominant influence on helminth transmission. Continuing environmental change will necessitate the adoption of mitigation and adaptation strategies in husbandry. Long term predictive models need to have the architecture to incorporate these changes. Ultimately, an optimal modelling approach is likely to combine mechanistic processes and physiological thresholds with correlative bioclimatic modelling, incorporating changes in livestock husbandry and disease control. Irrespective of approach, the principal limitation to parasite predictions is the availability of active surveillance data and empirical data on physiological responses to climate variables. By combining improved empirical data and refined models with a broad view of the livestock system, robust projections of helminth risk can be developed.

INTRODUCTION

Climate change has been implicated as a driving force for recent parasite range expansions, and efforts have been made to model the relationship between pathogen levels and climate. The most economically important parasitic helminths of livestock in temperate climes include the nematodes *Haemonchus contortus*, *Teladorsagia circumcincta* and *Nematodirus battus*, and the trematode *Fasciola hepatica*. The increase in these helminths in recent years [1,2,3,4,5] has been attributed to climate change, since the survival of the free-living stages is chiefly affected by temperature and moisture, and larval development rate is highly temperature dependent [6,7,8,9].

The development of evidence-based risk assessments and targeted surveillance are pivotal when the welfare and economic costs of these pathogens are considered. Subclinical infection is characterised by weight loss, lower milk yield, loss of condition, abortion, infertility and veterinary costs, and heavy infections can cause host mortality. Indirect economic losses from pathogen outbreaks are also incurred via export restrictions and surveillance and mitigation costs [10]. Despite the deleterious impacts of helminths on the livestock industry and their dependence on climatic conditions, predictions of long-term threats to animal health from climate change have so far concentrated on heat stress [11,12,13,14,15,16] and viruses spread by volant vectors, such as blue tongue [17,18,19,20,21,22]. Although there have been a number of studies aiming to link the recent changes in helminthiasis abundance and distribution with environmental change [1,3,4,5,23,24,25], there is a lack of predictions for future helminth risk to livestock.

Here we explore the optimal approaches to generating long-term predictions for helminth risk, stratifying modelling approaches as either correlative or mechanistic. We also highlight the obstacles to generating meaningful predictions and the need for a multidisciplinary approach.

MODELLING THE CHANGE

The Correlative Approach

Predictive models of species distribution, for both epidemiology and conservation, are often based on correlative ecological niche models [26,27,28,29,30,31,32]. These models are based on Hutchinson's ecological niche theory where the current geographic distribution is used to infer the environmental requirements of a species [33]. The current, past or future distribution of that species is then predicted based on these requirements [34]. Insights into the biology of parasite dynamics should be used to systematically build the foundations of these models, and the most proximal environmental predictors should be chosen based on known ecological and physiological theory [35].

A number of programmes are available to determine a species' climate envelope by matching current distribution with climatic parameters, such as CLIMEX [36], HABITAT [37], DOMAIN [38] and SPECIES [39]. In addition to these generic models, species-specific correlative models have also been developed, however these models have primarily been applied to species of conservation importance and invasive alien species. To date, correlative predictive models of helminthiasis have focussed on *F. hepatica* (liver fluke) due to the close relationship between weather and fasciolosis outbreaks [2] and the worldwide importance of fasciolosis as a zoonosis. The UK's National Animal Disease Information Service (NADIS) and the Department of Agriculture and Food in Ireland currently predict *F. hepatica* incidence using the Ollerenshaw index, which is a correlative model focusing on temperature and moisture availability [40,41]. Through applying a modified Ollerenshaw index to UKCP09 climate projections, the first long-term predictions of future *F. hepatica* risk for 2020–2070 across the UK have been developed [42]. The resultant risk maps predict unprecedented levels of fasciolosis outbreaks, and possible changes in the timing of disease outbreaks due to increased risk from overwintering larvae. Despite an overall long term increase in all regions of the UK, spatio-temporal variation in risk levels is expected, with infection risk due to reduce in some areas and fluctuate greatly in others. This forecast is the first approximation of the potential impacts of climate change on helminth risk in the UK, and indicates where active disease surveillance should be targeted.

There are a number of factors that made correlative modelling the preferred approach for determining fascioliasis risk. Firstly, the existence of long term prevalence data facilitated the development of a statistical model [41]. Secondly, despite their complex life cycle, the distribution of liver fluke is driven by simple proximal drivers—temperature and water availability. It is these drivers that form the basis of the correlative model. Thirdly, their dispersal by wild hosts and speed of colonisation of new regions makes them ubiquitous where livestock are present across their fundamental niche.

Despite the information that correlative models can provide, there are a number of disadvantages to applying this approach to climate change predictions. A prominent bottleneck to the development of correlative models is the lack of current or past parasite distribution data. The data used to train these models is usually opportunistic, passive surveillance data. Climate parameters are used according to which measurements and predictions are available, rather than those that are most pertinent to disease transmission. In order to build reliable models, purpose-driven active surveillance data is needed. As climate affects long term, large scale trends, these data must be collected over appropriate scales to observe these trends, rather than the patchy distribution visible at finer scales, which is likely to be a consequence of non-climatic factors. It is equally important to apply predictive models at the right scale. It is often tempting to apply models on as fine a scale as the climate projections allow, as this will lead to detailed maps with an illusion of greater accuracy. However this temptation should be resisted and models based on climatic parameters alone should only be applied at an appropriately coarse scale.

The lack of distribution data also impedes the validation of these predictive models, as ideally they should be validated with data fully independent from that used to build the model [26,28,43]. A further issue with validation is that it often indicates the models ability to predict current distribution; models that are effective at predicting current distribution may not be so reliable when predicting future outbreaks [34]. Hence there is a need to validate these models with data outside the spatial and temporal range of the training dataset. To make use of existing datasets for both model creation and validation there are ongoing developments in statistical methodology to account for biases in reported data. Existing datasets can be utilised for modelling distributions if imperfect detection is considered, with irregular sampling intensity and false absences accounted for using hierarchical Bayesian models [29,30,44,45]. Uncertainty in predictor variables can also be addressed through Bayesian analysis [46].

Given sufficient data for creation and validation of correlative models, fundamental disadvantages remain. As correlative models are generated using distribution data they are based on the realized niche [47] which includes competitive exclusion and biotic interactions (e.g., the presence of hosts and other pathogens), rather than the whole fundamental niche. As distribution data are usually only available for a limited area or time window, correlative models are generally fitted using data reflecting a snapshot of the current climate-pathogen relationship. An assumption is therefore made that a species has reached equilibrium with its environment [35]. This ignores any non-climatic dispersal constraints that may be restricting the current range, or declining sink populations where presence is recorded but climate is not actually suitable for long-term population survival. This failure to account for non-equilibrium ranges is an inherent weakness in correlative models [48,49]. This weakness could be especially pertinent for emerging parasites which are unlikely to be in equilibrium with their current environment, so could be absent from climatically suitable areas due to low propagule pressure. Conversely, these models will not identify areas which are not within bounds necessary for pathogen survival, but where deleterious levels of parasitism could be reached given sufficient propagule pressure. This issue posed by the non-equilibrium nature of emerging pathogens could be addressed through training models for emerging parasites using distribution data from their long established native range [35].

Above all aforementioned limitations, the dangers in extrapolating statistical models are especially pertinent to climate projections. The assumption that a species is in equilibrium with its current environment and the reliance on relationships between climate variables that may not exist in novel climate change scenarios leads to equivocal results when models are extrapolated spatially and temporally [29,44,50,51]. We are likely to experience previously unseen climate combinations and climate parameters exceeding their current observed ranges; parasites could be well adapted to these novel situations, however correlative models would assume these conditions to be unsuitable [34]. Additionally, correlative models cannot identify points where the system behaviour undergoes a qualitative change (tipping points). It is this inability to extrapolate that further emphasises the need for an alternative modelling approach.

As a further testament to correlative models' often unreliable outputs, models created using different statistical approaches have been shown to produce conflicting distribution estimates, even when trained with the same distribution and climate data [52,53]. Such differences between distribution estimates can be exacerbated when extrapolating to novel climatic conditions [32]. To address the divergence in model outputs, a framework for selecting the most robust statistical modelling technique has been developed [32]. This approach would be useful for selecting optimal models for predicting changes in helminth distribution. The differences between distributions predicted by different models could also be informative; discovering why predictions differ could improve understanding of the main drivers, and quantification of the differences could inform decision making and risk analysis [29].

Despite their limitations, correlative models can still provide a first indication of how climate will influence helminth distribution, and identify where limited resources and targeted surveillance should be focused. They provide a useful tool when too few elements of the transmission process have been quantified for the creation and parameterisation of mechanistic models. The limitations of correlative models are being addressed through continued development of statistical methodologies, led by work in both conservation and invasive species control; there is scope for these emerging approaches to be applied to predicting livestock parasite risk. Reliability of correlative models will ultimately be governed by the quality of data used for model training and validation, the statistical methods employed, the ecological and physiological knowledge on which inclusion of proximal variables is based, and discrepancies between the realized and fundamental niche.

The Mechanistic Approach

The second approach to understanding and predicting parasite risk is the process-based mechanistic approach. Mechanistic models are based on detailed knowledge of the physiology of the species [34] and attempt to replicate the underlying mechanisms that drive the species' response to environmental variables [54]. Given sufficient understanding of the parasites physiology, these models can be employed to predict changes in prevalence [55]. Previous mechanistic models have explored the dynamics of helminth infection in livestock [5 6,57,58,59,60,61,62,63,64,65]. However these models have not been used to assess the impacts of climate change on helminth transmission.

One element of mechanistic modelling that makes it well adapted in a changing climate is that it is less prone to extrapolation problems than correlative models. The mechanistic approach does not rely on relationships between climate variables that may cease to exist under future climate change [34,47,66,67], making them less prone to breaking down when tested outside current observation limits. Consequently, the mechanistic approach is considered superior in extrapolating beyond current conditions and forecasting the impacts of climate change [68]. When predicting a parasite's response to novel conditions it is vital that models account for the possibility of non-linear responses to predictor variables [43]. Through developing a mechanistic model looking at the effects of increasing temperatures on Schistosomiasis prevalence and abundance, Mangal *et al.* (2008) found that *Schistosoma mansoni* showed a non-linear response to temperature change [68]. Extrapolated predictions for novel climates may prove to be unreliable if the response curves are not fully determined, hence empirical data on survival and development under conditions beyond a parasite's current climatic ranges are required.

The mechanistic approach also avoids some of the inherent problems that correlative models have when applied to emerging and invasive parasites since mechanistic models do not assume the species is in equilibrium with the environment [48] and are capable of predicting tipping points. By incorporating key mechanisms into the model structure more complex questions can also be addressed, such as determining the impact of climate on specific stages of the lifecycle.

The severity of helminth infections is often dependent on intensity, rather than prevalence. The economic and welfare implications of outbreaks would be difficult to evaluate using a correlative approach, as data availability typically constrains these models to only look at prevalence. If data on adult worm burden are available they are often based on indirect measures such as faecal egg count. It is more feasible to include models of adult worm burden within a mechanistic framework. Although mechanistic models can predict changing intensity, the relationship between helminth burden and host production losses is non-linear; a doubling of infection intensity does not lead to a doubling of production loss [69]. A deeper understanding of the relationship between parasite burdens and potential production losses is still required.

The survival of larvae on pasture is largely driven by climate [62], but helminth population dynamics are governed by density dependent process and it is vital that these processes are incorporated (e.g., host acquired immunity), as more larvae on pasture do not necessarily translate to higher parasite burden in the livestock [64]. Previous parasite models have concentrated on infection dynamics in a single host, returning the average host infection level and assuming each host interacts with its own private population of parasites [57,59,62,64]. In reality, aggregation of parasites in host populations is a ubiquitous phenomenon [70]. It has been proposed that heterogeneities in host infection levels arise due to numerous factors including variations in host resistance and susceptibility due to behavioural or physiological differences and spatial heterogeneity in infection risk [71,72,73,74]. Average host models ignore this overdispersion of helminths; a potentially critical feature in a system heavily driven by density-dependent processes.

By modelling the average host, it is assumed that the same level of regulatory constraint is experienced by all parasites in the population. In an overdispersed population, aggregation results in an increase in the effective density experienced by individual parasites, increasing the influence of regulatory processes [74,75,76]. These regulatory processes drive a parasite population towards equilibrium [63], and hence aggregated populations reach equilibrium more quickly and at a lower level. As a consequence, models that do not take the overdispersed parasite distribution into account are likely to overestimate burden [75,76]. Variation in host worm burden within a herd has been generated artificially in infection models through assuming a fixed distribution of worms across hosts, and the system having a pre-defined degree of overdispersion [56,75]. There is scope to develop models where an overdispersed distribution is an emergent property, and which subsume both the mechanisms that generate aggregation and the impacts of the resultant aggregation on infection dynamics. Future changes in grazing season, stocking density, host behaviour and host resistance may affect the aggregation of parasites at the suprapopulation level (defined as all parasites of a given species, at all stages of development, within an ecosystem [77]). Models that incorporate the changing distribution of the suprapopulation would also allow identification of scenarios where individuals within a herd are at risk from heavy infections, thus addressing economic and welfare concerns.

Unlike correlative models, mechanistic model development is not restricted by the paucity of accurate distribution data. Instead, the primary restriction to their development is the extensive physiological data needed for parameterisation [35]. Recent empirical work on how livestock parasites respond to changing climatic variables should facilitate their development and parameterisation. For example, the correlation between the timing, distribution and quantity of rainfall and *H. contortus* development have been investigated [6] and threshold rates of precipitation and evaporation required for survival of *H. contortus* larvae to the infective stage have been determined [78]. The impact of temperature on the emergence of the infective stages of trematodes from the intermediate snail hosts have also been reviewed [79,80]. These studies demonstrate that there are quantitative data on the effects of climatic variables on physiological processes, which can be used to parameterise mechanistic models.

In addition to altered temperature and precipitation, changes in UV reaching the extra-mammalian stages will also vary with changing climate (due to e.g., varying cloud cover, changes in development window and seasonality of infection). Van Dijk *et al.* (2009) investigated how ultraviolet (UV) light affects Trichostrongyloid nematodes, *H. contortus*, *N. battus* and *T. circumcinta*, in their infective L3 stages [81]. Mortality rates increased by up to 100 times due to exposure to UVA light equivalent to the maximum levels expected for a summers day. There were inter-species differences observed: *N. battus*, with its arctic origins, and *T. circumcincta* were less resistant than the tropic-adapted *H. contortus* [81]. Previous mechanistic models have focused on changing temperature, as this was the parameter most studied and available in climate projections. However this study highlights the importance of including other environmental variables in predictive models.

Another oversight in some mechanistic models is that they assume Liebig's Law of the Minimum [34], which presumes that the overall response will be determined by the most limiting factor. This is not necessarily true. For example, larvae on pasture could survive otherwise deleteriously low levels of rainfall if temperatures were not too high. This emphasises the need to look at proximal, rather than distal variables—in this case water availability rather than levels of rainfall. It also emphasises the importance of looking at variables in combination.

In contrast with the correlative approach, mechanistic models are based on measured physiological and behavioural parameters, and reflect the fundamental rather than realized niche [35]. This leads to over estimation of risk showing the whole potential range if dispersal restrictions or biotic interactions are not accounted for [34,48,82,83]. The extent of predictive errors will be partly dependent on the proportion of the fundamental niche that the realized niche occupies [35]. This proportion will depend on the parasites dispersal and competitive abilities; if the discrepancies between the species' fundamental and realized niche could be identified, it could give an indication of how accurate predictive models could be for particular species.

To attain realistic predictions of species distribution, models should ultimately integrate constraints from biotic interactions and dispersal [35]. Projections often assume either complete dispersal, where the parasite can reach all areas with permissive climes, or no dispersal where presence is only predicted in areas of its current range where climate remains within viable thresholds [35]. It is overly simplistic to assume that a species will either fully disperse to fill the fundamental niche, or that they will only survive in areas where the projected fundamental niche and current realized niche overlap [26], however, this approach does at least provide (often very wide) bounds on the possible distribution. The importance of incorporating species interactions in prediction models has been demonstrated [51,84,85], with biotic interactions influencing the predictive power of correlative models even at the macro-ecological scale [85]. The incorporation of dispersal also influences predictive power [30,31,86], with the disparity between models that do incorporate dispersal and those that do not increasing under more extreme climate warming scenarios [31]. The integration between model projections, biotic interactions and simulated host/pathogen distribution and dispersal patterns would improve the accuracy of predictive models (both correlative and mechanistic). This could be achieved through the adoption of hierarchical Bayesian frameworks [86]. However, these should take into account that biotic interactions and dispersal abilities could change with time. Due to the high selective pressures that novel conditions are likely to impose, climate change is likely to select for phenotypes with enhanced dispersal capabilities that can track the shift in their climate envelope. This would affect the parameterisation of models that incorporate dispersal.

Mechanistic models provide a powerful tool in predicting the influence of climate change on helminth risk. They are comparatively robust under spatio-temporal extrapolation and can be developed to answer complex questions. However, a move towards a mechanistic approach should not be seen as a way of alleviating the need for distribution data; to develop and validate models and assess their continued reliability under changing conditions, there is a need for ongoing surveillance.

LOOKING BEYOND CLIMATE: THE NEED FOR A PANOPTIC VIEW

Climate is an important driver of helminth distribution, however transmission to livestock at the farm level is also controlled by husbandry. Livestock husbandry is ever changing and the adoption of mitigation and adaptation strategies will be necessary under continuing environmental change. Methods for alleviating heat stress in cattle have been explored including shade provision [14] and different feeding and sprinkling regimes [87]. Changes in the time of highest feed production may alter, resulting in timing of livestock reproduction being managed differently [88]. Housing outdoors instead of indoors in winter will also affect parasite transmission cycles [16]. Additionally, as certain breeds are more susceptible to extreme weather conditions, the breed structure of livestock may have to change to ensure that production levels are maintained and animal welfare needs met.

The influence of these adaptation and mitigation strategies on parasite risk has yet to be incorporated in pathogen transmission models. There is a propensity for predictive models to incorporate ever refined climate prediction, however these will not provide meaningful indications of future risk if they are parameterised for outdated management approaches and redundant host population structures. Predictive models need to have the architecture to allow adaptation and mitigation strategies to be incorporated. For example, as different breeds have varying levels of disease susceptibility [10,13,89], the differing levels of resistance for future breed structures need to be considered in transmission models.

The architecture of mechanistic models should also allow more complex issues to be addressed, for example, determining which of the panoply of potential control strategies resources should be invested into developing. There is currently a strong reliance on anthelmintics for controlling helminths in intensively grazed livestock [90], either through targeting the number of adult parasites in the host, or the number of infectious larvae on pasture. However development of anthelmintic resistance is an ongoing problem [5,23,91,92,93,94] and resistance development could be accelerated in a warming climate [90]. As anthelmintics become decreasingly efficient in combating these parasites, strategies to lower the dependence on chemicals and manage the spread of resistance need adopting [95,96,97,98,99]. There are a variety of alternative control strategies, often linked with organic production. These include grazing management strategies to decrease the number of infective larvae that the host is exposed to, and breeding for resistance in the host [100,101,102,103]. Models should be developed with the architecture to consider interacting effects of changes in climate, management and control strategies.

CONCLUSIONS AND RECOMMENDATIONS

Although correlative modelling is a useful tool for establishing baseline predictions, a drive towards mechanistic process-based models will ultimately be needed if we are to foresee the consequences of subtle interactions between various components of a system under climate influence. Although laboratory studies have been done to aid the creation of mechanistic models, the findings have not yet been utilised. To improve the reliability of extrapolating and applying models to novel climates, empirical evidence is also needed on how helminths respond when exposed to conditions beyond their current observed ranges, to account for any non-linearity.

Modelling approaches can be broadly stratified as correlative or mechanistic, but in practice it is difficult to draw an absolute distinction between these modelling techniques. Ultimately, an integration of both modelling approaches may be required. Future improvements in predictions should arise from the continued development of a hybrid approach that combines both mechanistic processes and physiological thresholds with correlative bioclimatic modelling. [28,50,51,104]. Irrespective of the modelling approach, the quality of predictions is critically dependent on the quality of available data [105]. The continued collection of active surveillance data and empirical data on physiological responses to climate variables will ultimately drive the development and validation of meaningful predictions.

The recent expansion of helminth ranges has been mirrored by our increased understanding of the role of climate in their transmission dynamics. However, climate is just one factor affecting disease ecology. The application of mitigation and adaptation strategies, combined with changing control options, need to be considered when determining the overall impacts of climate change. By combining improved empirical data and refined models with a broader view of the livestock system, projections of future disease threats can be improved. This will increase our understanding of these complex systems, identifying potential risks and highlighting opportunities for control.

ACKNOWLEDGEMENTS

SAC and BioSS receive funding from the Scottish Government

REFERENCES

1. Mitchell, G.B.B.; Somerville, D.K. Effects of Climate Change on Helminth Diseases in Scotland. *SAC Publication* 2005, *1*, 1–11.
2. de Waal, T.; Relf, V.; Good, B.; Gray, J.; Murphy, T.; Forbes, A.; Mulcahy, G. Developing models for the predicion of fasciolosis in Ireland. In *Making Science Work on the Farm: A Workshop on Decision Support Systems for Irish Agriculture*; Holden, N.M., Hochstrasser, T., Schulte, R.P.O., Walsh, S., Eds.; Joint Working Group in Applied Agricultural Meteorology: Dublin, Ireland.
3. Kenyon, F.; Sargison, N.D.; Skuce, P.J.; Jackson, F. Sheep helminth parasitic disease in south eastern Scotland arising as a possible consequence of climate change. *Vet. Parasitol.* 2009, *163*, 293–297.
4. Pritchard, G.C.; Forbes, A.B.; Williams, D.J.L.; Salimi-Bejestani, M.R.; Daniel, R.G. Emergence of fasciolosis in cattle in East Anglia. *Vet. Record* 2005, *157*, 578–582.
5. van Dijk, J.; David, G.P.; Baird, G.; Morgan, E.R. Back to the future: Developing hypotheses on the effects of climate change on ovine parasitic gastroenteritis from historical data. *Vet. Parasitol.* 2008, *158*, 73–84.
6. O'Connor, L.J.; Kahn, L.P.; Walkden-Brown, S.W. Moisture requirements for the free-living development of *Haemonchus contortus*: Quantitative and temporal effects under conditions of low evaporation. *Vet. Parasitol.* 2007, *150*, 128–138.
7. Coyne, M.J.; Smith, G. The development and mortality of the free-living stages of *Haemonchus contortus* in laboratory culture. *Int. J. Parasitol.* 1992, *22*, 641–650.

8. Barnes, E.H.; Dobson, R.J.; Donald, A.D.; Waller, P.J. Predicting populations of *Trichostrongylus colubriformis* infective larvae on pasture from meteorological data. *Int. J. Parasitol.* 1988, *18*, 767–774.

9. Armour, J. The epidemiology of helminth disease in farm animals. *Vet. Parasitol.* 1980, *6*, 7–46.

10. Wilson, A.; Mellor, P. Bluetongue in Europe: Vectors, epidemiology and climate change. *Parasitol. Res.* 2008, *103*, 69–77.

11. Beatty, D.; Barnes, A.; Taylor, E.; Maloney, S. Do changes in feed intake or ambient temperature cause changes in cattle rumen temperature relative to core temperature? *J. Therm. Biol.* 2008, *33*, 12–19.

12. García-Ispierto, I.; López-Gatius, F.; Bech-Sabat, G.; Santolaria, P.; Yániz, J.L.; Nogareda, C.; De Rensis, F.; López-Béjar, M. Climate factors affecting conception rate of high producing dairy cows in northeastern Spain. *Theriogenology* 2007, *67*, 1379–1385.

13. Harle, K.; Howden, S.; Hunt, L.; Dunlop, M. The potential impact of climate change on the Australian wool industry by 2030. *Agr. Syst.* 2007, *93*, 61–89.

14. Kendall, P.E.; Nielsen, P.P.; Webster, J.R.; Verkerk, G.A.; Littlejohn, R.P.; Matthews, L.R. The effects of providing shade to lactating dairy cows in a temperate climate. *Livestock Sci.* 2006, *103*, 148–157.

15. Nardone, A.; Ronchi, B.; Lacetera, N.; Bernabucci, U. Climatic Effects on Productive Traits in Livestock. *Vet. Res. Commun.* 2006, *30*, 75–81.

16. Gregory, N.G. How climatic changes could affect meat quality. *Food Res. Int.* 2010, *43*, 1866–1873.

17. Caligiuri, V.; Giuliano, G.A.; Vitale, V.; Chiavacci, L.; Travaglio, S.; Salute, V. Bluetongue surveillance in the Campania Region of Italy using a geographic information system to create risk maps. *Epidemiol. Vect.* 2004, *40*, 385–389.

18. de Koeijer, A.; Hartemink, N.; Boender, G.J.; Elbers, A.; Heesterbeek, H. Epidemiological analysis of the 2006 bluetongue virus serotype 8 epidemic in north-western Europe A risk map for epidemic potential in The Netherlands. *Eur. Food Safety Author.* 2007, *1*, 1–18.

19. Gubbins, S.; Carpenter, S.; Baylis, M.; Wood, J.L.N.; Mellor, P.S. Assessing the risk of bluetongue to UK livestock: Uncertainty and sensitivity analyses of a temperature-dependent model for the basic reproduction number. *J. Roy. Soc. Interf.* 2008, *5*, 363–371.

20. Pili, E.; Ciucce, S.; Culurgioni, J.; Figus, V.; Pinna, G.; Marchi, A. Distribution and abundance of Bluetongue vectors in Sardinia: Comparison of field data with prediction maps. *J. Vet. Med.* 2006, *53*, 312–316.

21. Purse, B.V.; Tatem, A.J.; Caracappa, S.; Rogers, D.J.; Mellow, P.S.; Baylis, M.; Torina, A. Modelling the distributions of *Culicoides* bluetongue virus vectors in Sicily in relation to satellite-derived climate variables. *Med. Vet. Entomol.* 2004, *18*, 90–101.

22. Racloz, V.; Presi, P.; Vounatsou, P.; Schwermer, H. Use of mapping and statistical modelling for the prediction of bluetongue occurrence in Switzerland based on vector biology. *Vet. Ital.* 2007, *43*, 513–518.

23. Abbott, K.A.; Taylor, P.M.; Stubbings, L.A. *Sustainable Worm Control Strategies for Sheep*, 2nd ed.; A Technical Manual for Veterinary Surgeons and Advisers; SCOPS (Sustainable Control of Parasites in Sheep),National Sheep Association: Malvern, UK, 2007; pp. 1–48.

24. Mas-Coma, S.; Valero, M.A.A.; Bargues, M.D. Effects of climate change on animal and zoonotic helminthiases. *Rev. Sci. Tech. Off. Int. Epiz* 2008, *27*, 443–457.

25. O'Connor, L.J.; Walkden-Brown, S.W.; Kahn, L.P. Ecology of the free-living stages of major trichostrongylid parasites of sheep. *Vet. Parasitol.* 2006, *142*, 1–15.

26. Lawler, J.J.; White, D.; Neilson, R.P.; Blaustein, A.R. Predicting climate-induced range shifts: Model differences and model reliability. *Global Change Biol.* 2006, *12*, 1568–1584.

27. Leathwick, J.; Elith, J.; Hastie, T. Comparative performance of generalized additive models and multivariate adaptive regression splines for statistical modelling of species distributions. *Ecol. Mode.* 2006, *199*, 188–196.

28. Heikkinen, R.K.; Luoto, M.; Araújo, M.B.; Virkkala, R.; Thuiller, W.; Sykes, M.T. Methods and uncertainties in bioclimatic envelope modelling under climate change. *Progr. Phys. Geogr.* 2006, *30*, 1–27.

29. Elith, J.; Leathwick, J.R. Species distribution models: Ecological explanation and prediction across space and time. *Ann. Rev. Ecol. Evol. Syst.* 2009, *40*, 677–697.
30. Pagel, J.; Schurr, F.M. Forecasting species ranges by statistical estimation of ecological niches and spatial population dynamics. *Global Ecol. Biogeogr.* 2012, *21*, 293–304.
31. Engler, R.; Guisan, A. MigClim: Predicting plant distribution and dispersal in a changing climate. *Div. Distrib.* 2009, *15*, 590–601.
32. Thuiller, W. BIOMOD—Optimizing predictions of species distributions and projecting potential future shifts under global change. *Global Change Biol.* 2003, *9*, 1353–1362.
33. Hutchinson, G.E. Concluding remarks. *Cold Spring Harb. Symp. Quant. Biol.* 1957, *22*, 415–427.
34. Hijmans, R.J.; Graham, C.H. The ability of climate envelope models to predict the effect of climate change on species distributions. *Global Change Biol.* 2005, *12*, 2272–2281.
35. Guisan, A.; Thuiller, W. Predicting species distribution: Offering more than simple habitat models. *Ecol. Lett.* 2005, *8*, 993–1009.
36. Sutherst, R.W.; Maywald, G.F. A computerised system for matching climates in ecology. *Agr. Ecosyst. Environ.* 1985, *13*, 281–299.
37. Walker, P.A.; Cocks, K.D. HABITAT: A procedure for modelling a disjoint environmental envelope for a plant or animal species. *Global Ecol. Biogeogr. Lett.* 1991, *1*, 108–118.
38. Carpenter, G.; Gillison, A.N.; Winter, J. DOMAIN: A flexible modelling procedure for mapping potential distributions of plants and animals. *Biodiv. Conserv.* 1993, *2*, 667–680.
39. Pearson, R.G.; Dawson, T.P.; Berry, P.M.; Harrison, P.A. SPECIES: A spatial evaluation of climate impact on the envelope of species. *Ecol. Model.* 2002, *154*, 289–300.
40. Ollerenshaw, C.B. The approach to forecasting the incidence of fascioliasis over England and Wales 1958-1962. *Agr. Meteorol.* 1964, *3*, 35–53.
41. Ollerenshaw, C.B.; Rowlands, W. A method of forecasting the incidence of fascioliasis in Anglesey. *Vet. Record* 1959, *71*, 591–598.

42. Fox, N.J.; White, P.C.L.; McClean, C.J.; Marion, G.; Evans, A.; Hutchings, M.R. Predicting impacts of climate change on *Fasciola hepatica* risk. *PloS One* 2011, *6*, e16126.
43. Austin, M. Species distribution models and ecological theory: A critical assessment and some possible new approaches. *Ecol. Model.* 2007, *200*, 1–19.
44. Elith, J.; Graham, C.H.; Anderson, R.P.; Dudik, M.; Ferrier, S.; Guisan, A.; Hijmans, R.J.; Huettmann, F.; Leathwick, J.R.; Lehmann, A.; Li, J.; Lohmann, L.G.; Loiselle, B.; Manion, G.; Moritz, C.; Nakamura, M.; Nakazawa, Y.; Overton, J.M.; Peterson, A.T.; Phillips, S.J.; Richardson, K.; Scachetti-Pereira, R.; Schapire, R.E.; Soberon, J.; Williams, S.; Wisz, M.S.; Zimmermann, N.E. Novel methods improve prediction of species' distributions from occurrence data. *Ecography* 2006, *29*, 129–151.
45. Bierman, S.M.; Butler, A.; Marion, G.; Kühn, I. Bayesian image restoration models for combining expert knowledge on recording activity with species distribution data. *Ecography* 2010, *33*, 451–460.
46. McInerny, G.J.; Purves, D.W. Fine-scale environmental variation in species distribution modelling: Regression dilution, latent variables and neighbourly advice. *Meth. Ecol. Evol.* 2011, *2*, 248–257.
47. Guisan, A.; Zimmermann, N.E. Predictive habitat distribution models in ecology. *Ecol. Model.* 2000, *135*, 147–186.
48. Pearson, R.G.; Dawson, T.P. Predicting the impacts of climate change on the distribution of species: Are bioclimate envelope models useful? *Global Ecol. Biogeogr.* 2003, *12*, 361–371.
49. Sutherst, R.W.; Bourne, A.S. Modelling non-equilibrium distributions of invasive species: A tale of two modelling paradigms. *Biol. Invas.* 2008, *11*, 1231–1237.
50. Elith, J.; Kearney, M.; Phillips, S. The art of modelling range-shifting species. *Meth. Ecol. Evol.* 2010, *4*, 330–342.
51. Kearney, M.; Porter, W. Mechanistic niche modelling: Combining physiological and spatial data to predict species' ranges. *Ecol. Lett.* 2009, *12*, 334–350.
52. Ortega-Huerta, M.A.; Peterson, A.T. Modeling ecological niches and predicting geographic distributions: A test of six presence-

only methods. *Revista Mexicana De Biodiversidad* 2008, *79*, 205–216.

53. Pearson, R.G.; Thuiller, W.; Arau, M.B.; McClean, C.; Miles, L. Model-based uncertainty in species range prediction. *J. Biogeogr.* 2006, *33*, 1704–1711.
54. Robertson, M.P.; Peter, C.I.; Villet, M.H.; Ripley, B.S. Comparing models for predicting species' potential distributions: A case study using correlative and mechanistic predictive modelling techniques. *Ecol. Model.* 2003, *164*, 153–167.
55. Anderson, R.M. The role of mathematical models in helminth population biology. *Int. J. Parasitol.* 1987, *17*, 519–529.
56. Barnes, E.H.; Dobson, R.J. Population dynamics of *Trichostrongylus colubriformis* in sheep: Computer model to simulate grazing systems and the evolution of anthelmintic resistance. *Int. J. Parasitol.* 1990, *20*, 823–831.
57. Callinan, A.P.L.; Morley, F.H.W.; Arundel, J.H.; White, D.H. A model of the life cycle of sheep nematodes and the epidemiology of nematodiasis in sheep. *Agr. Syst.* 1982, *9*, 199–225.
58. Cornell, S.J.; Isham, V.S.; Grenfell, B.T. Stochastic and spatial dynamics of nematode parasites in farmed ruminants. *Proc. Roy. Soc. B* 2004, *271*, 1243–1250.
59. Dobson, R.J.; Donald, A.D.; Barnes, E.H.; Waller, P.J. Population dynamics of *Trichostrongylus colubriformis* in sheep: Model to predict the worm population over time as a function of infection rate and host age. *Int. J. Parasitol.* 1990, *20*, 365–373.
60. Gordon, G.; O'Callaghan, M.; Tallis, G.M. A Deterministic Model for the Life Cycle of a Class of Internal Parasites of Sheep. *Math. Biosci.* 1970, *8*, 209–226.
61. Learmount, J.; Taylor, M.A.; Smith, G.; Morgan, C. A computer model to simulate control of parasitic gastroenteritis in sheep on UK farms. *Vet. Parasitol.* 2006, *142*, 312–329.
62. Leathwick, D.M.; Barlow, N.D.; Vlassoff, A. A model for nematodiasis in New Zealand lambs. *Int. J. Parasitol.* 1992, *22*, 789–799.
63. Paton, G.; Thomas, R.G.; Waller, P.J. A prediction model for parasitic gastro-enteritis in lambs. *Int. J. Parasitol.* 1984, *14*, 439–445.

64. Roberts, M.G.; Grenfell, B.T. The Population Dynamics of Nematode Infections of Ruminants: Periodic Perturbations as a Model for Management. *J. Math. Appl. Med. Biol.* 1991, *8*, 83–93.

65. Tallis, G.M.; Donald, A.D. Further models for the distribution on pasture of infective larvae of the strongylid nematode parasites of sheep. *Math. Biosci.* 1970, *7*, 179–190.

66. Dormann, C.F. Promising the future? Global change projections of species distributions. *Basic Appl. Ecol.* 2007, *8*, 387–397.

67. Kearney, M.R.; Wintle, B.A.; Porter, W.P. Correlative and mechanistic models of species distribution provide congruent forecasts under climate change. *Conserv. Lett.* 2010, *3*, 203–213.

68. Mangal, T.D.; Paterson, S.; Fenton, A. Predicting the impact of long-term temperature changes on the epidemiology and control of Schistosomiasis: A mechanistic model. *PLoS One* 2008, *1*, e1438.

69. Smith, G. The economics of parasite control: Obstacles to creating reliable models. *Vet. Parasitol.* 1997, *72*, 437–449.

70. Gregory, R.D.; Woolhouse, M.E. Quantification of parasite aggregation: A simulation study. *Acta tropica* 1993, *54*, 131–139.

71. Barbour, A.D.; Kafetzaki, M. Modeling the Overdispersion of Parasite Loads. *Math. Biosci.* 1991, *107*, 249–253.

72. Cornell, S.J.; Isham, V.S.; Smith, G.; Grenfell, B.T. Spatial parasite transmission, drug resistance, and the spread of rare genes. *PNAS* 2003, *100*, 7401–7405.

73. Louie, K.; Vlassoff, A.; Mackay, A. Nematode parasites of sheep: Extension of a simple model to include host variability. *Parasitology* 2005, *130*, 437–446.

74. Shaw, D.J.; Dobson, A.P. Patterns of macroparasite abundance and aggregation in wildlife populations: A quantitative review. *Parasitology* 1995, *111*, 111–133.

75. Churcher, T.S.; Ferguson, N.M.; Basáñez, M.G. Density dependence and overdispersion in the transmission of helminth parasites. *Parasitology* 2005, *131*, 121–132.

76. Smith, G.; Guerrero, J. Mathematical models for the population biology of *Ostertagia ostertagi* and the significance of aggregated parasite distributions. *Vet. Parasitol.* 1993, *46*, 243–257.

77. Esch, G.W.; Gibbons, J.W.; Bourque, J.E. An analysis of the relationship between stress and parasitism. *Am. Midland Naturalist* 1975, *93*, 339–353.
78. O'Connor, L.J.; Kahn, L.P.; Walkden-Brown, S.W. Interaction between the effects of evaporation rate and amount of simulated rainfall on development of the free-living stages of *Haemonchus contortus*. *Vet. Parasitol.* 2008, *155*, 223–234.
79. Poulin, R. Global warming and temperature-mediated increases in cercarial emergence in trematode parasites. *Parasitology* 2006, *132*, 143–151.
80. Mas-Coma, S.; Valero, M.A.; Bargues, M.D. Climate change effects on trematodiases, with emphasis on zoonotic fascioliasis and schistosomiasis. *Vet. Parasitol.* 2009, *163*, 264–280.
81. van Dijk, J.; de Louw, M.D.E.; Kalis, L.P.A.; Morgan, E.R. Ultraviolet light increases mortality of nematode larvae and can explain patterns of larval availability at pasture. *Int. J. Parasitol.* 2009, *39*, 1151–1156.
82. Aurambout, J.P.; Finlay, K.J.; Luck, J.; Beattie, G.A.C. A concept model to estimate the potential distribution of the Asiatic citrus psyllid (*Diaphorina citri* Kuwayama) in Australia under climate change—A means for assessing biosecurity risk. *Ecol. Model.* 2009, *220*, 2512–2524.
83. Lafferty, K.D. The ecology of climate change and infectious diseases. *Ecology* 2009, *90*, 888–900.
84. Davis, A.J.; Jenkinson, L.S.; Lawton, J.H.; Shorrocks, B.; Wood, S. Making mistakes when predicting shifts in species range in response to global warming. *Nature Lett.* 1998, *391*, 783–786.
85. Araújo, M.B.; Luoto, M. The importance of biotic interactions for modelling species distributions under climate change. *Global Ecol. Biogeogr.* 2007, *16*, 743–753.
86. Hooten, M.B.; Wikle, C.K.; Dorazio, R.M.; Royle, J.A. Hierarchical spatiotemporal matrix models for characterizing invasions. *Biometrics* 2007, *63*, 558–567.
87. Mader, T.L.; Davis, M.S. Effect of management strategies on reducing heat stress of feedlot cattle: Feed and water intake. *J. Anim. Sci.* 2004, *82*, 3077–3087.

88. Rivington, M.; Matthews, K.B.; Bellocchi, G.; Buchan, K.; Stöckle, C.O.; Donatelli, M. An integrated assessment approach to conduct analyses of climate change impacts on whole-farm systems. *Environ. Model. Softw.* 2007, *22*, 202–210.

89. Eguale, T.; Mekonnen, G.A.; Chaka, H. Evaluation of variation in susceptibility of three Ethiopian sheep breeds to experimental infection with *Fasciola hepatica*. *Small Rumin. Res.* 2009, *82*, 7–12.

90. Wolstenholme, A.J.; Fairweather, I.; Prichard, R.; Samson-Himmelstjerna, G.V.; Sangster, N.C. Drug resistance in veterinary helminths. *Trends Parasitol.* 2004, *20*, 469–476.

91. DeVaney, J.A.; Craig, T.A.; Rowe, L.D. Resistance to ivermectin by *Haemonchus contortus* in goats and calves. *Int. J. Parasitol.* 1992, *22*, 369–376.

92. Le Jambre, L.F.; Southcott, W.H.; Dash, K.M. Resistance of selected lines *of Haemonchus contortus* to Thiabendazole, Morantel, Tartrate and Levamisole. *Int. J. Parasitol.* 1976, *6*, 217–222.

93. Pandey, V.S.; Sivaraj, S. Anthelmintic resistance in *Haemonchus contortus* from sheep in Malaysia. *Vet. Parasitol.* 1994, *53*, 67–74.

94. Yadav, C.L.; Kumar, R.; Uppal, R.P.; Verma, S.P. Multiple anthelmintic resistance in *Haemonchus contortus* on a sheep farm in India. *Vet. Parasitol.* 1995, *60*, 355–360.

95. Ketzis, J.K.; Vercruysse, J.; Stromberg, B.E.; Larsen, M.; Athanasiadou, S.; Houdijk, J.G.M. Evaluation of efficacy expectations for novel and non-chemical helminth control strategies in ruminants. *Vet. Parasitol.* 2006, *139*, 321–335.

96. Torres-Acosta, J.F.J.; Hoste, H. Alternative or improved methods to limit gastro-intestinal parasitism in grazing sheep and goats. *Small Rumin. Res.* 2008, *77*, 159–173.

97. Waller, P.J. From discovery to development: Current industry perspectives for the development of novel methods of helminth control in livestock. *Vet. Parasitol.* 2006, *139*, 1–14.

98. Waller, P.J. Sustainable nematode parasite control strategies for ruminant livestock by grazing management and biological control. *Anim. Feed Sci. Technol.* 2006, *126*, 277–289.

99. Barger, I.A. Prospects for integration of novel parasite control options into grazing systems. *Int. J. Parasitol.* 1996, *26*, 1001–1007.

100. Woolaston, R.; Barger, I.A.; Piper, L.R. Response to helminth infection of sheep selected for resistance to *Haemonchus contortus*. *Int. J. Parasitol.* 1990, *20*, 1015–1018.

101. Gicheha, M.; Kosgey, I.; Bebe, B.; Kahi, A. Efficiency of alternative schemes breeding for resistance to gastrointestinal helminths in meat sheep. *Small Rumin. Res.* 2007, *69*, 167–179.

102. Windon, R.G. Genetic control of resistance to helminths in sheep. *Vet. Immun. Immunopathol.* 1996, *54*, 245–54.

103. Woolaston, R.R.; Baker, R.L. Prospects of breeding small ruminants for resistance to internal parasites. *Int. J. Parasitol.* 1996, *26*, 845–55.

104. Monahan, W.B. A mechanistic niche model for measuring species' distributional responses to seasonal temperature gradients. *PloS One* 2009, *4*, e7921.

105. Rocchini, D.; Hortal, J.; Lengyel, S.; Lobo, J. M.; Jiménez-Valverde, A.; Ricotta, C.; Bacaro, G.; Chiarucci, A. Accounting for uncertainty when mapping species distributions: The need for maps of ignorance. *Progr. Phys. Geogr.* 2011, *35*, 211–226.

Chapter 4

Climate Change, Crop Production and Child Under Nutrition in Ethiopia: A Longitudinal Panel Study

Seifu Hagos[1,3], Torleif Lunde[3], Damen H Mariam[1], Tassew Woldehanna[2], and Bernt Lindtjørn[3]

[1]School of Public health, College of Health sciences, Addis Ababa University, Addis Ababa, Ethiopia

[2]Department of Economics, College of Business and economics, Addis Ababa University, Addis Ababa, Ethiopia

[3]Center for International Health, University of Bergen, Bergen, Norway

ABSTRACT

Background

The amount and distribution of rainfall and temperature influences household food availability, thus increasing the risk of child under nutrition. However, few studies examined the local spatial variability and the impact of temperature and rainfall on child under nutrition at a smaller scale (resolution). We conducted this study to evaluate the effect of weather variables on child under nutrition and the variations in effects across the three agro ecologies of Ethiopia.

Methods

A longitudinal panel study was conducted. We used crop productions (cereals and oilseeds), livestock, monthly rainfall and temperature, and child under nutrition data for the period of 1996, 1998, 2000 and 2004. We applied panel regression fixed effects model.

Results

The study included 43 clusters (administrative zones) and 145 observations. We observed a spatio temporal variability of rainfall, stunting and underweight. We estimated that for a given zone, one standard deviation increase in rainfall leads to 0.242 standard deviations increase in moderate stunting. Additionally, a one standard deviation increase temperature leads to 0.216 standard deviations decrease in moderate stunting. However, wasting was found to be poorly related with rainfall and temperature. But severe wasting showed a positive relationship with the quadratic term of rainfall.

Conclusions

We conclude that rainfall and temperature are partly predicting the variation in child stunting and underweight. Models vary in predicting stunting and underweight across the three agro ecologic zones. This could indicate that a single model for the three agro ecologies may not be not applicable.

BACKGROUND

Ethiopia has experienced repeated famine since the 9th century [1]. Drought due to failure of rains often precedes Ethiopian famines. The failure of rain results in crop failure, impact food productions and usually results in food shortages in vulnerable parts of the population. Historical accounts showed that famine declines after the arrival of rains [2]. Rainfall is hence, one of the most important factors influencing livelihoods of subsistence farmers and pastoralists. Failures or irregularities of the rainy season have a direct link to reduced household food availability [3]. Therefore, in some parts of the country the pattern of rainfall during the main growing season of June-July-August-September (JJAS) has grave consequences on crop availability and child nutrition.

The spatio temporal distribution as well as the amount of rain and temperature influence human health[4]. The influence is substantial in developing countries, such as Ethiopia, which are largely dependent on rain fed agriculture [5]. Climate change impacts food security through multiple pathways. These include altering the availability of food that depends on the agricultural production[6] and influencing the stability of food supplies due to extreme weather events. Moreover, climate impacts are observed through influencing access to food and utilization [7].

There is a marked improvement in children's anthropometric status in Ethiopia over the past 10 years, as seen by the downward trend in the proportion of children stunted and underweight over the three successive Ethiopian Demographic and Health Survey (EDHS) [8-10]. Although the trend is showing a decreasing pattern, the proportion of under-five stunting, underweight and wasting are still high and more efforts are needed to reach the MDG goals [11]. Climate change is one of the challenges against the efforts undergoing to combat child under nutrition through improved household food security. Ethiopia recently (1999–2000) experienced the effect of low and untimely rainfall [12].

Children are vulnerable to the effects of climate change and examples of these effects are reviewed and reported [13-18]. However, few studies examined the local spatial variability and the impact of climate on stunting, underweight and wasting of under-five children despite the fact that children are considered vulnerable. The purpose of this study was to: (i) evaluate the spatial distribution of rainfall, temperature, per capita crop availability and under nutrition; (ii) characterize the pattern and interrelationship of rainfall, temperature and under nutrition; and (iii) analyze the variations in effects across different agro ecological zones of Ethiopia.

Hence this paper builds on the links between climate variables and under-five children under nutrition and could be used to design appropriate programs for areas impacted by climate change.

METHODS

Study Design and Period

We employed a longitudinal panel study design to estimate the effect of growing season temperature and rainfall on child under nutrition for the period of 1996–2004.

Data and Data Sources

The dataset constituted a panel of observations of multiple variables. Crop productions (cereals and oil seeds) and livestock data for administrative zones included in the study were obtained from the Central Statistics Agency (CSA Ethiopia) for the period of 1996, 1998, 2000, and 2004. We converted the total amount of crops produced during the main harvesting season (October to November) in each zone into per capita crop availability. We used projected population for each administrative zone. The projected population of zones was estimated by CSA using the ratio method based on the projected population of each region.

We obtained monthly rainfall data for the months of JJAS for each studied zones from the respective weather station(s). The data were made available by the Ethiopian Malaria Prediction System Research Project and the National Meteorological Authority. The main harvest season in most of the study locations is during the months of October and November. We assumed that crop yield is predominantly affected by the amount of rain during the growing seasons of JJAS. Hence we computed the total amount of rainfall for the pre harvest seasons of JJAS and used in this analysis.

The main outcomes of interest for this study were both moderate and severe forms of stunting, wasting, and underweight in children under five years of age. Children were considered moderately malnourished if one of the three forms of under nutrition are 2 SDs (standard deviations) below the median expected height-for-age, weight-for-height and weight-for-age. Children were considered severely malnourished if one of the three forms of under nutrition is 3 SDs below the median expected height-for-age, weight-for-height and weight-for-age.

We used data sets of Agricultural Sample and the Demographic Health Survey (DHS) Surveys collected by the Central Statistics Agency (CSA) that cover all Ethiopian administrative zones from 1996 to 2004. Out of these successive data sets, we created a pseudo panel data set of under nutrition, crop, livestock and other variables at zonal level. The zonal level panel data sets of under nutrition, crop, livestock and other data were matched by year.

Altitude data at 30 arc-seconds (~1 km) resolution data were downloaded by a tile from the world climate data source website (http://www.worldclim.org/bioclim *webcite*.). The altitudes were then extracted from the raster data set for each study zone. Temperature data were obtained from the Climatic Research Unit (CRU) TS (time-series) datasets (CRU v321) available athttp://badc.nerc.ac.uk/view/badc.nerc.ac.uk__ATOM__ACTIVITY_0c08abfc-f2d5-11e2-a948-00163e251233

Data Processing and Analysis

We used Stata (version 11, Stata Corporation, College Station, TX) for panel data analysis. Spatial visualization, extraction of altitude data and mapping was done using Arc GIS version 10 (ESRI).

We applied panel data regression techniques and used the variables per capita crop arability, livestock, rainfall (both linear and quadratic) and temperature in the model to estimate the effect of climate variability on child wasting, underweight and stunting.

Before applying regression, multiple steps were followed. We took logs of crop per capital availability to achieve normality. Based on altitude, we classified the study areas (administrative zones) into three agro ecological zones. A separate model was fit to see the variations in response to climate across the three agro ecological zones. We calculated standardized anomalies for all the variables and used the same in the model. By standardizing we ask if wetter/warmer conditions in any zone leads to more/less under nutrition in the same zone.

Hausman test was conducted in order to choose between fixed or random effects models. The test basically examines whether the error terms are correlated with the regressors. The null hypothesis was that the preferred model is a random effect, while the alternative is the fixed effects. If the error terms are correlated with the one or more of the regressors (such as rainfall), the estimated coefficients are biased and hence the preferred model is the fixed effects.

We used both the linear and quadratic terms for standardized rainfall assuming that stunting, underweight and wasting could be worsened by extreme low and high rainfall. However, we did not check for serial correlation of the residuals as the data set constructed had a shorter timer series.

Goodness of Fits of Models

We reported three types of R-square values for each regression model. These are the within, between, and overall R-squares. The within R-square value indicates how much of the variation in child under nutrition with in a zone over the study period is explained by weather variables. The between R-square values indicates how much of the variation in child under nutrition between zones is explained by weather variables. The overall R-square values indicates how much of the overall variation in child under nutrition is explained by climatic variables.

Non-Technical Summary of the Methods

We constructed a panel dataset using the following steps. For each study zone, data on rainfall, temperature, per capita crop availability and livestock was compiled for the years 1996, 1998, 2000 and 2004. We then matched child under nutrition estimates of each zone with the respective study year. This made a panel data set consisting of a total of 145 observations. A panel data set consists of a cross sectional time series data in which attributes (e.g. Rainfall) of many units (e.g., Zones) are observed over time (e.g. Years). From this data we computed standardized anomalies in order to capture the effects of climate within a given zone. The data were then analyzed using a fixed effects model. We chose the fixed effects model over random effects because each study zone can have a peculiar feature or characteristics that can prevent (e.g. higher productivity) or worsens child nutrition condition and this must be accounted in the analysis.

RESULTS

Sample Characteristics

The study included 43 clusters (administrative zones) and 145 observations for the period of 1996–2004. This period was one of the recent periods that Ethiopia experienced the effect of low and untimely rainfall [12]. There were on average 3.5 observations per cluster (zone) in the data set. A descriptive summary of the panel data set used for the present analysis is presented (Table 1).

Table 1: Summary of panel data used in the study, Ethiopia, 1996–2004

Variable		Mean	Std. dev.	Min	Max	Observations*
Rainfall (mm)	Overall	645.2	317.4	41.2	1377.9	N=145
	Between		306.5	64.4	1225.4	n=41
	Within		122.3	258.8	1055.0	T-bar=3.5
Temperature (°C)	Overall	19.92	3.12	15.5	29.9	N=142
	Between		3.56	15.8	29.9	n=41
	Within		0.19	19.44	20.3	T-bar=3.6
Per capita crop (kg)	Overall	206.9	166.9	10.8	1022.3	N=127
	Between		123.6	22.9	528.6	n=34
	Within		115.5	-106.5	872.6	T-bar=3.73529
Wasting (%)	Overall	10.1	3.7	2.8	24.6	N=145
	Between		2.9	3.4	17.9	n=41
	Within		2.6	2.7	16.8	T-bar=3.5
Severe wasting (%)	Overall	3.3	2.1	0.0	16.3	N=143
	Between		1.5	0.0	9.2	n=41
	Within		1.6	-0.7	10.4	T-bar=3.4

Underweight (%)	Overall	42.5	9.5	19.0	62.5	N=145
	Between		7.4	27.6	57.9	n=41
	Within		6.3	23.5	57.7	T-bar=3.5
Severe underweight (%)	Overall	15.4	5.6	3.4	30.3	N=145
	Between		4.1	7.5	21.8	n=41
	Within		4.0	5.2	24.9	T-bar=3.5
Stunting (%)	Overall	55.2	11.2	20.4	78.6	N=145
	Between		7.4	38.3	70.6	n=41
	Within		8.8	28.2	73.8	T-bar=3.5
Severe stunting	Overall	32.8	10.0	8.0	54.2	N=145
	Between		6.8	18.5	45.9	n=41
	Within		7.5	15.4	50.9	T-bar=3.5

*N=total number of observations, n=the number of clusters (zones), T-bar=average observation per cluster (zone).

Hagos *et al.*

Hagos *et al. BMC Public Health* 2014 14:884, mdoi:10.1186/1471-2458-14-884

Summary of the Model Parameters

The average growing season rainfall of the study locations was 645.2 mm. The overall amount of growing season rainfall ranges from 41.2 to 1378 mm. The average minimum and maximum amount of rainfall with in zones was 64.4 and 1225.4 mm respectively. However, the average minimum and maximum amount of rainfall documented between zones was 258 and 1055 mm respectively. We observed a comparatively higher variability of rainfall between zones compared to within zones.

We found an average growing season temperature of 19.9°C for the study period. The overall minimum and maximum temperatures for the study locations were 15.5°C and 29.9°C respectively. Unlike rainfall, we observed a smaller variability of temperature within zones over the study period.

Low amount of rainfall during the growing season is expected to affect crop production and availability. Data on total per capita crop availability in the studied zones indicated that the overall mean total per capita crop availability during the study period was 206.9 kg. Per capita crop availability in the study ranged from 10.8 kg to 1022.3 kg.

The overall average prevalence of underweight over the study period was 42.5% (range: 19, 62.5%). About 15.4% had a severe degree of underweight. The overall average stunting during the study period was 55.2% (range: 20.4, 78.6%). About 32.8% of the children had severe stunting. The mean wasting prevalence over the study period was 10.1%. We observed a higher variability in the prevalence of moderate stunting, wasting and underweight. The variability was consistent between as well as within the zones over the study period.

Spatial and Temporal Pattern

The average growing season rainfall shows marked variation over the study period as well as agro-ecologies. The average growing season rainfall showed a decreasing pattern between the highlands to the lowlands. The highlands had an average rainfall of 726.8 mm, while the midlands and lowlands had 668.0 mm and 513.9 mm respectively. The average growing season temperature didn't show a marked variation over the study period. We observed a relatively higher temperature in lowland compared to the midlands and highlands (Table 2).

Table 2: Distribution of rainfall, temperature, per capita crop and child under nutrition status by agro ecologies and study period, Ethiopia, 1996–2004

Agro ecology	Year	Rainfall (mm) (mean, SD)	Temperature (°C) (mean, SD)	Per capita crop (kg)	wasting (%) (mean, SD)		Underweight (%) (mean, SD)		Stunting (%) (mean, SD)	
					Moderate	Severe	Moderate	Severe	Moderate	Severe
Lowland	1996	547.3(335.9)	21.42.0)	391.1(210.5)	8.1(2.7)	3.9(0.7)	41.6(7.9)	16.7(5.3)	63.1(6.3)	42.4(8.3)
	1998	540.8(401.3)	23.1(3.0)	130.6(121.8)	9.5(3.8)	2.9(1.8)	41.0(8.0)	14.0(4.9)	55.8(7.6)	32.2(6.5)
	2000	512.0(414.7)	23.6(3.4)	162.5(154.8)	11.2(4.5)	3.7(3.9)	38.4(9.8)	13.9(6.6)	47.6(9.7)	27.0(8.0)
	2004	464.6(374.0)	22.5(2.3)	138.2(136.3)	9.8(5.4)	2.8(2.5)	34.7(10.1)	12.4(6.9)	41.7(10.5)	21.3(7.9)
	Total	513.9(377.2)	22.8(2.6)	202.3(185.1)	9.9(4.3)	3.3(2.6)	38.79(9.2)	14.0(6.0)	51.0(11.5)	29.6(10.3)
Midlands										
	1996	721.1(165.4)	18.5(2.3)	259.4(129.1)	9.7(2.6)	4,2(1.1)	43.0(6.3)	17.9(4.6)	65.0(6.0)	43.7(6.8)
	1998	751.8(304.1)	19.7(2.7)	187.3(197.6)	11.6(4.4)	2.9(1.4)	48.4(8.0)	16.3(4.9)	57.3(7.4)	32.8(7.2)
	2000	646.4(263.5)	19.5(2.9)	224.8(167.3)	12.0(2.5)	4.7(2.2)	45.7(10.7)	18.6(5.7)	56.8(7.6)	33.4(7.3)
	2004	645.3(292.9)	19.7(2.8)	278.2(271.8)	10.8(4.2)	3.1(1.8)	35.7(4.6)	10.3(3.1)	43.9(10.9)	22.2(7.7)
	Total	688.0(281.9)	19.4(2.7)	233.9(198.0)	11.2(3.6)	3.7(1.9)	43.2(9.1)	15.6(5.7)	54.8(11.0)	32.2(9.8)
Highlands										
	1996	733.0(165.4)	17.5(1.4)	217.1(160.6)	9.6(2.5)	4.1(1.5)	45.9(7.6)	19.5(5.3)	64.7(6.4)	43.9(7.4)
	1998	744.2(194.5)	18.1(1.5)	163.5(111.9)	9.9(3.3)	2.3(1.0)	49.3(8.0)	15.3(4.1)	62.3(8.7)	35.0(7.3)
	2000	739.1(340.3)	17.9(1.6)	186.9(100.2)	10.5(2.7)	3.3(1.5)	38.4(9.8)	18.9(3.6)	61.8(6.7)	37.8(6.80
	2004	691.2(274.0)	18.1(1.5)	182.2(271.8)	7.0(2.6)	2.1(1.2)	36.1(7.5)	12.8(3.7)	49.5(9.4)	28.6(7.8)
	Total	726.8(247.3)	17.9(1.5)	186.99118.0)	9.26(3.0)	2.9(1.5)	45.3(9.1)	16.6(4.9)	59.5(9.8)	36.2(9.0)
All zones		645.2(317,4)	19.92(3.11)	206.9(167.0)	10.1(3.7)	3.3(2.1)	42.5(9.5)	15.4(5.6)	55.2(11.2)	32.8(10.0)

Hagos *et al.*

Hagos *et al. BMC Public Health* 2014 14:884, doi:10.1186/1471-2458-14-884

We observed a decreasing pattern on stunting, wasting and underweight over the study periods. The highlands and midlands documented a relatively higher prevalence of stunting and underweight compared to the lowlands. The average prevalence of underweight and stunting among highlands was 45.3 and 59.5% respectively, while the average prevalence of underweight and stunting among lowlands was 38.7 and 51.0% respectively. However, the prevalence of severe forms of stunting, underweight and wasting did not vary over the three agro ecologic zones (Table 2).Figure 1 shows the spatial pattern of average growing season rainfall, per capita crop availability, stunting and underweight in the study zones. The choropleth map indicates zones in the north and northwestern part of the country documented higher rainfall. Higher per capita crop availability is documented in some zones with high rainfall. We noted that zones with higher rainfall documented a higher prevalence of stunting and underweight compared with zones with lower rainfall. However, zones with higher per capita crop availability had a relatively lower prevalence of child under nutrition rates.

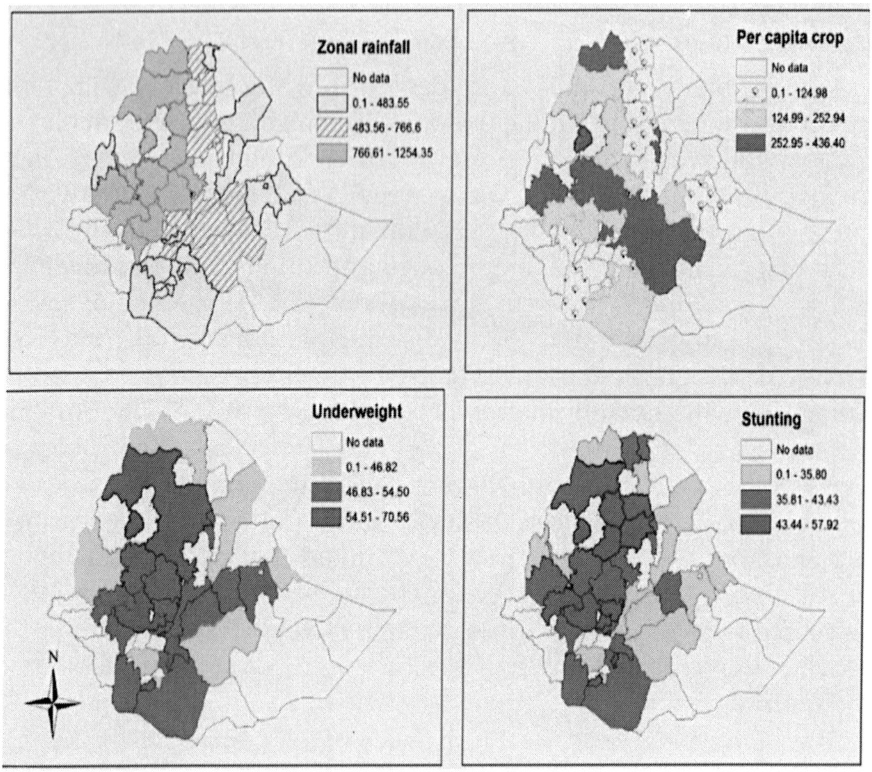

Figure 1: Choropleth maps showing spatial distribution of rainfall, per capita crop availability and under nutrition for the study locations, Ethiopia,1996-2004.

Panel Regression Results

We observed the following relationships from the panel regression models for moderate and severe stunting (Tables 3 and 4). First, stunting was found to be strongly and negatively correlated with growing season temperature for the three agro ecologies. For a given zone, one standard deviation increase in temperature resulted in 0.216 standard deviation decrease in moderate stunting. This relationship is statistically significant for the all zone and the lowland models. Second, stunting is positively associated with the amount of rainfall, indicating that an increase in rainfall resulting in an increas. For a given zone, one standard deviation increase in rainfall resulted in 0.242 standard deviation increase in moderate stunting. This relationship is statistically significant for the all zone and midlands model. We did not find any significant result on the relationship between rainfall and stunting when the quadratic terms instead of the linear form of rainfall was used. However, the direction of the coefficients indicates that extreme forms rainfall is leading to a higher prevalence of moderate stunting. Similar results were documented on the relationship between severe form of stunting and weather variables.

Table 3: Panel regression results on the effect of rainfall and temperature on stunting, Ethiopia, 1996–2004

	All zone model		Lowlands		Midlands		Highlands	
	coef†	se‡	coef	se	coef	se	coef	se
Rainfall during growing season	0.242***	0.91	0.06	0.17	0.495**	0.16	0.22	0.16
Temperature	-0.216*	0.12	-0.55**	0.26	-0.09	0.17	-0.23	0.22
Quadratic term for rainfall	0.060	0.14	-0.33	0.27	0.30	0.22	0.23	0.27
Per capita crop	-0.53	0.000	0.003	0.18	0.12	0.16	-0.000	0.17
Livestock	-0.02*	0.068	-0.27*	0.24	-0.52	0.21**	0.06	0.18
Number of observations	121		32		39		50	
Within R2	0.28		0.53		0.49		019	
Between R2	0.81		0.07		0.95		0.87	
Overall R2	0.30		0.53		0.5		0.19	

Note: ***p<0.01, **p<0.05, *p<0.1 , †model coefficients, ‡standard errors of the coefficient.

Hagos *et al.*

Hagos *et al. BMC Public Health* 2014 14:884, doi:10.1186/1471-2458-14-884

Table 4: Panel regression results on the effect of rainfall and temperature on severe stunting, Ethiopia, 1996–2004

	All zone model		Lowlands		Midlands		Highlands	
	coef†	se‡	coef	se	coef	se	coef	se

Rainfall during growing season	0.225***	0.81	-0.01	0.17	0.51**	0.16	0.17	0.13
Temperature	-0.24*	0.11	-0.64**	0.26	-0.28	0.18	-0.13	0.17
Quadratic term for rainfall	0.52	0.12	-0.39	0.27	0.26	0.22	0.27	0.21
Per capita crop	0.13	0.85	0.99	0.18	0.23	0.16	0.13	0.13
Livestock	-0.39***	0.10	-0.15	0.22	-0.36	0.21**	-0.46***	0.15
Number of observations	121		32		39		50	
Within R2	0.43		0.56		0.50		0.47	
Between R2	0.82		0.03		0.81		0.97	
Overall R2	0.44		0.56		0.51		0.48	

Note: ***p<0.01, **p<0.05, *p<0.1 , †model coefficients, ‡standard errors of the coefficient.

Hagos *et al.*

Hagos *et al. BMC Public Health* 2014 14:884, doi:10.1186/1471-2458-14-884

Tables 5 and 6 shows results from the models where moderate and severe underweight is regressed on growing season rainfall and temperature. The results across the agro ecological zones consistently showed inverse relationships between growing season temperature and moderate form of underweight, although the relationship was not statistically significant. A severe form of underweight, however, showed a statistically significant inverse association with temperature. We observed that the quadratic term for rainfall is significantly related with underweight in the highland models indicating that a very small as well as high amount of rainfall leading to higher prevalence of underweight. This relation might indicate a nonlinear relationship between rainfall and underweight in the highlands of Ethiopia.

Table 5: Panel regression results on the effect of rainfall and temperature on underweight, Ethiopia, 1996–2004

	All zone model		Lowlands		Midlands		Highlands	
	coef†	se‡	coef	se	coef	se	coef	se

Rainfall during growing season	0.19*	0.10	-0.17	0.21	0.36*	0.20	0.23	0.16
Temperature	-0.12	0.14	-0.29	0.34	0.07	0.23	-0.21	0.23
Quadratic term for rainfall	0.12	0.16	-0.32	0.35	0.21	0.27	0.49*	0.26
Per capita crop	-0.07	0.11	0.10	0.23	-0.21	0.20	-0.05	0.16
Livestock	-0.08	0.12	-0.16	0.28	-0.31	0.24	0.15	0.18
Number of observations	121		32		39		50	
Within R2	0.09		0.22		0.22		0.25	
Between R2	0.80		0.03		0.45		0.25	
Overall R2	0.09		0.22		0.22		0.25	

Note: *p<0.1 , †model coefficients, ‡standard errors of the coefficient.

Hagos *et al.*

Hagos *et al. BMC Public Health* 2014 14:884, doi:10.1186/1471-2458-14-884

Table 6: Panel regression results on the effect of rainfall and temperature on severe underweight, Ethiopia, 1996–2004

	All zone model		Lowlands		Midlands		Highlands	
	coef†	se‡	coef	se	coef	se	coef	se
Rainfall during growing season	0.14	0.10	-0.27	0.20	0.42**	0.18	0.19	0.15
Temperature	-0.26**	0.13	-0.35	0.30	-0.26	0.21	-0.32	0.17
Quadratic term for rainfall	0.16	0.15	-0.39	0.31	0.02	0.25	-0.03	0.24
Per capita crop	0.17*	0.01	0.37*	0.21	0.14	0.18	0.14	0.15
Livestock	-0.18	0.12	0.09	0.25	-0.27	0.22	-0.27	0.17
Number of observations	121		32		39		50	
Within R2	0.24		0.36		0.35		0.35	
Between R2	0.06		0.01		0.07		0.97	
Overall R2	0.24		0.36		0.34		0.35	

Note: **p<0.05, *p<0.1 , †model coefficients, ‡standard errors of the coefficient.

Hagos *et al.*

Hagos *et al. BMC Public Health* 2014 14:884, doi:10.1186/1471-2458-14-884

In the present study wasting was found to be poorly related with rainfall and temperature (Table 7). None of the models resulted a significant relationship with rainfall and temperature. But severe wasting showed a positive relationship with the quadratic term of rainfall in the all zone as well as the midland models (Table 8).

Table 7: Panel regression results on the effect of rainfall and temperature on wasting, Ethiopia, 1996–2004

	All zone model		Lowlands		Midlands		Highlands	
	coef†	se‡	coef	se	coef	se	coef	se
Rainfall during growing season	-0.06	0.10	-0.04	0.17	0.01	0.20	-0.03	0.18
Temperature	-0.14	0.14	0.33	0.26	-0.32	0.24	-0.21	0.23
Quadratic term for rainfall	0.09	0.16	-0.37	0.27	0.31	0.28	0.37	0.29
Per capita crop	-0.04	0.11	0.33	0.18	-0.22	0.20	-0.02	0.18
Livestock	0.11	0.13	0.16	0.22	0.15	0.25	-0.04	0.20
Number of observations	121		32		39		50	
Within R2	0.02		0.2		0.50		0.47	
Between R2	0.17		0.00		0.81		0.97	
Overall R2	0.02		0.2		0.51		0.48	

Note:, †model coefficients, ‡standard errors of the coefficient.

Hagos *et al.*

Hagos *et al. BMC Public Health* 2014 14:884, doi:10.1186/1471-2458-14-884

Table 8: Panel regression results on the effect of rainfall and temperature on severe wasting, Ethiopia, 1996–2004

	All zone model		Lowlands		Midlands		Highlands	
	coef†	se‡	coef	se	coef	se	coef	se
Rainfall during growing season	0.04	0.10	0.09	0.18	0.29	0.20	-0.14	0.15

Temperature	-0.11	0.13	0.05	0.28	-0.17	0.23	-0.11	0.2
Quadratic term for rainfall	0.26*	0.14	0.19	0.29	0.49*	0.26	0.26	0.25
Per capita crop	0.26**	0.1	0.62	0.19**	0.09	0.19	0.23	0.15
Livestock	-0.26**	0.12	-0.10	0.23	-0.32	0.23	-0.41**	0.17
Number of observations	121		32		39		50	
Within R2	0.27		0.45		0.22		0.35	
Between R2	0.05		0.02		0.45		0.91	
Overall R2	0.26		0.45		0.22		0.35	

Note: **p<0.05, *p<0.1 , †model coefficients, ‡standard errors of the coefficient.

Hagos *et al.*

Hagos *et al. BMC Public Health* 2014 14:884, doi:10.1186/1471-2458-14-884

Goodness of Fits of Models

Except for the lowlands, the variation in moderate and severe stunting between study zones is adequately explained by climatic variables in the model (between R square values; 0.81-0.95). However, the variation in stunting with a study zone over the study period is poorly explained by the present model. The R-square values for models on underweight were very small and vary over the three agro ecologies. The overall R-square values indicated that these models capture smaller aspects of the variation in child underweight between as well as within the study zones over the study period.

DISCUSSION

We used existing data to explore spatio temporal patterns and further to estimate the impact of growing season temperature and rainfall on child underweight, wasting and stunting before, during and after the crisis period of 1999–2000 [12].We found that unlike temperature, rainfall showed a marked variation over the study periods as well as agro-ecologies. Although, there is a decreasing pattern of stunting, wasting and underweight over time , a higher prevalence of stunting and underweight were found in the highlands and midlands compared to the lowlands. We found that the amount and direction of the effect of rainfall vary among the different ecologies. Additionally, sometimes the quadratic terms of rainfall rather than the linear forms were significant predictors for underweight and stunting. Moreover, the results of the study demonstrate that temperature has a significant effect on child underweight and stunting.

The reported prevalence of stunting, wasting, and underweight from this study are relatively higher when compared with successive EDHS [8-10]. However, a similar result was observed with regard to the decreasing trend in stunting and underweight over the study years. Moreover when the prevalence of stunting and underweight of the present study is compared with EDHS survey of the same year, we found a comparable figure All these taken might suggest that the sample could be represent significantly major parts of Ethiopia.

A similar approach was used to provide evidence on the association between climate change and child under nutrition in Mali, Africa. [19]. Stunting is found to be highly influenced by arid climate even when controlled for livelihoods. However, the effect of climate on underweight is found to be not significant. Some argue that underweight is a short term response to climate seasonal flux or shocks and these shocks can be absorbed and modified by livelihood adaptation capabilities. Moreover, sometimes the effects of decline in rainfall (and crop failure) on child hood anthropometries indices may not be visible as it could be prevented through public health measures [20]. Our study also documented that the model for underweight showed non-significant association with rainfall as compared to models for stunting.

The finding that rainfall and temperature predicting child stunting has important implications over future child under nutrition attributable to climate change. A relative increase on moderate and severe forms of stunting is estimated due to climate change in sub-Saharan African countries [21]. However, uncertainties still remain on the pattern of future rainfall in Eastern African countries including Ethiopia. Studies done by Christensen et al. reported a higher probability of an increase in the annual mean rainfall in East Africa extending to the Horn of Africa [22]. The growing seasons of countries such as Ethiopia would be benefited due to a combination of increased rainfall as well as temperature indicating that not all changes in climate variability would be negative [23]. In contrary, Funk et al.[24] indicated that warming of the Indian Ocean would lead to a decrease in rainfall and hence can threaten Eastern Africa. Some argue that the precipitation simulation by IPPC did not consider the complex terrain nature of the Eastern Africa [24].

Assessing the effect of climate variability on health poses methodological challenges. The common challenges include exposure assessment, ecological fallacies, the complexity of relationships, and scale of the study [25,26]. First, in this study we assumed that the exposure (rainfall and temperature) will be similar for households that are found in the same zone (group) as climate impacts populations rather than individuals [25]. Hence we interpreted the link between rainfall, temperature and child under nutrition at zonal (group) level using aggregated estimates. However, in the absence of individually collected data, it is somehow difficult to exclude totally the role of ecological fallacy in the relationship. Moreover, we cannot rule out the local variation in the exposures such as rainfall and temperature within a given zone. Second, we used the UNICEF conceptual framework [27] to develop a biologically plausible model. Immediate and underlying causes of child under nutrition are captured with model variable such illness prevalence, livestock and per capita crop availability. However, the actual relationship can be more complex than assumed, and can be nonlinear requiring multiple pathways.

The findings of the current study shall be interpreted within the context of the following limitations. Due to the limitation of the availability of complete data for such work, the sample sizes for a stratified analysis based on agro-ecological zones were small. This has likely resulted in the absence of significant results. We were not also able to quantify and characterize threshold limits of rainfall which would have been beneficial to model child under nutrition risks. Despite these limitations, we believe that the present study involved more than half of the administrative zones of Ethiopia and was able to generate important information on the variation in effects of weather variables on child under nutrition.

CONCLUSIONS

We conclude that rainfall and temperature are partly predicting the variation in stunting and underweight in Ethiopia. Moreover, the models vary in predicting stunting and underweight across the three agro ecologic zones. This could indicate that a single model for all the three agro ecologies may not be not applicable. We recommend further work but at a micro level using similar analysis methods to assess the effect of rainfall and temperature on stunting, wasting and underweight.

AUTHORS' CONTRIBUTIONS

SH conceived the study, collected, analyzed and prepared the draft manuscript (as partial requirement for his PhD degree at University of Bergen, Norway). TL helped to analyze and interpret the data. BL, as a primary advisor for SH, conceived the study, advised and supervised the conception, the conduct as well as the analysis and write-up of the study. DH advised and supervised the analysis and write-up of the study. TW advised and supervised the conception, the analysis and write-up of the study. All authors helped in drafting the manuscript. All authors have read and approved the manuscript.

ACKNOWLEDGEMENTS

We would like to thank the Ethiopian Central Statistics Agency (CSA Ethiopia) for providing data on crop production, cattle production and population size. We would also like to thank the National Meteorological Authority (NMA) for providing station based rainfall data.

REFERENCES

1. Taye A, Mariam DH, Murray V: Interim report: review of evidence of the health impact of famine in Ethiopia. *Perspect Public Health* 2010, 130:222-226.
2. Kloss H, Lindtjorn B: Malnutrition during recent famines in Ethiopia. *Northeast Afr Stud* 1994, 1:121-136.
3. Haile M: Weather patterns, food security and humanitarian response in sub-Saharan Africa. *Philos Trans R Soc Lond B Biol Sci* 2005, 360:2169-2182.

4. Campbell-Lendrum D, Woodruff R: Comparative risk assessment of the burden of disease from climate change. *Environ Health Perspect* 2006, 114:1935-1941.
5. Parrya ML, Rosenzweigb C, Iglesiasc A, Livermored M, Fischere G: Effects of climate change on global food production under SRES emissions and socio-economic scenarios. *Glob Environ Chang* 2004, 14:53-67.
6. Lobell DB, Schlenker W, Costa-Roberts J: Climate trends and global crop production since 1980. *Science* 2011, 333:616-620.
7. Confalonieri U, Menne B, Akhtar R, Ebi KL, Hauengue M, Kovats RS, Revich B, Woodward A:Human health. In *Climate Change 2007: Impacts, Adaptation and Vulnerability. Contribution of Working Group II to the Fourth Assessment Report of the Intergovernmental Panel on Climate Change*. Cambridge, UK: Cambridge University Press; 2007:391-431.
8. Central Statistical Agency [Ethiopia] and ORC Macro: *Ethiopia Demographic and Health Survey 2000*. Addis Ababa, Ethiopia and Calverton, Maryland, USA: Central Statistical Agency and ORC Macro; 2001. Ref Type: Report
9. Central Statistical Agency [Ethiopia] and ORC Macro: *Ethiopia Demographic and Health Survey 2005*. Addis Ababa, Ethiopia and Calverton, Maryland, USA: Central Statistical Agency and ORC Macro; 2006. Ref Type: Report
10. Central Statistical Agency [Ethiopia] all: *Ethiopia Demographic and Health Survey 2011*. Addis Ababa, Ethiopia and Calverton, Maryland, USA: Central Statistical Agency and ORC Macro; 2012. Ref Type: Report
11. Stevens GA, Finucane MM, Paciorek CJ, Flaxman SR, White RA, Donner AJ, Ezzati M: Trends in mild, moderate, and severe stunting and underweight, and progress towards MDG 1 in 141 developing countries: a systematic analysis of population representative data. *Lancet* 2012, 380:824-834.
12. Hammond L, Maxwell D: The Ethiopian crisis of 1999–2000: lessons learned, questions unanswered. *Disasters* 2002, 26:262-279.
13. Bernstein AS, Myers SS: Climate change and children's health. *Curr Opin Pediatr* 2011, 23:221-226.

14. Ebi KL, Paulson JA: Climate change and children. *Pediatr Clin North Am* 2007, 54:213-226.Vii
15. Ebi KL, Paulson JA: Climate change and child health in the United States. *Curr Probl Pediatr Adolesc Health Care* 2010, 40:2-18.
16. Etzel RA: Climate change and child health in the United States: foreword. *Curr Probl Pediatr Adolesc Health Care* 2010, 40:1.
17. McCartney PR: Climate change and child health. *MCN Am J Matern Child Nurs* 2007, 32:255.
18. Seal A, Vasudevan C: Climate change and child health. *Arch Dis Child* 2011, 96:1162-1166.
19. Jankowskaa MM, Lopez-Carrb D, Funkc C, Husakd GJ, Chafee ZA: Climate change and human health: Spatial modeling of water availability, malnutrition, and livelihoods in Mali, Africa. *Appl Geogr* 2011, 33:4-15.
20. Kumar HR, Venkaiah K, Arlappa N, Kumar S, Brahmam GNV, Vijayaraghavan K: Diet and Nutritional Situation of the Population in the Severely Drought Affected Areas of Gujarat. *J Hum Ecol* 2005, 18(4):319-326.
21. Lloyd SJ, Kovats RS, Chalabi Z: Climate change, crop yields, and undernutrition: development of a model to quantify the impact of climate scenarios on child undernutrition. *Environ Health Perspect* 2011, 119:1817-1823.
22. Christensen JH, Hewitson B, Busuioc A, Chen A, Gao X, Held I, Jones R, Kolli RK, Kwon W-T, Laprise R, Magaña Rueda V, Mearns L, Menéndez CG, Räisänen J, Rinke A, Sarr A, Whetton P:Regional Climate Projections. In *Climate Change 2007: The Physical Science Basis. Contribution of Working Group I to the Fourth Assessment Report of the Intergovernmental Panel on Climate Change*. Edited by Solomon S, Qin D, Manning M, Chen Z, Marquis M, Averyt KB, Tignor M, Miller HL. Cambridge, United Kingdom and New York, NY, USA: Cambridge University Press; 2007.
23. Thornton PK, Jones PG, Owiyo TM, Kruska RL, Herero M, Kristjanson P, Notenbaert A, Bekele N, Omolo A: Mapping Climate Vulnerability and Poverty in Africa. In *Report to the Department for International Development*. Nairobi: ILRI; 2006:200. Ref Type: Report

24. Funk C, Dettinger MD, Michaelsen JC, Verdin JP, Brown ME, Barlow M, Hoell A: Warming of the Indian Ocean threatens eastern and southern African food security but could be mitigated by agricultural development. *Proc Natl Acad Sci U S A* 2008, 105:11081-11086.

25. Xun WW, Khan AE, Michael E, Vineis P: Climate change epidemiology: methodological challenges. *Int J Public Health* 2010, 55:85-96.

26. Ebi KL: Healthy people 2100: modeling population health impacts of climate change. *Clim Change* 2008, 88:5-19.

27. UNICEF: *Strategy for Improved Nutrition of Children and Women in Developing Countries: A UNICEF Policy Review*. New York; 1990:5-36. Ref Type: Report

Chapter 5

Herdsmen's Adaptation to Climate Changes and Subsequent Impacts in the Ecologically Fragile Zone, China

Yingcheng Liu[1], Tao Zhang[2,3], Xiaoli Geng[1], Liansheng He[4], and Zhiguo Pang[5,6]

[1]Faculty of Resources and Environmental Science, Hubei University, Wuhan, Hubei 430062, China

[2]Institute of Geographic Sciences and Natural Resources Research, Chinese Academy of Sciences, Beijing 100101, China

[3]Center for Chinese Agricultural Policy, Chinese Academy of Sciences, Beijing 100101, China [4]State Key Laboratory of Environmental Criteria and Risk Assessment, Chinese Research Academy of Environmental Sciences (CRAES), Beijing 100012, China

[5]State Key Laboratory of Simulation and Regulation of Water Cycle in River Basin, Beijing 100048, China

[6]China Institute of Water Resources and Hydropower Research, Beijing 100048, China

ABSTRACT

The change of land surface can exert significant influence on the future climate change. This study analyzed the effects of herdsmen's adaptation to climate changes on the livestock breeding, income, and land surface dynamics with a land surface parameterization scheme. The empirical analysis was first carried out on the impacts of the adaptation measures of herdsmen on their income in the context of the climate change with the positive mathematical programming (PMP) model on the basis of the household survey data in the Three-River Source Region, an ecologically fragile area in Qinghai Province, China. Then, the land surface parameterization process is analyzed based on the agent-based model (ABM), which involves the herdsmen's adaptation measures on climate change, and it also provides reference for the land surface change projection. The result shows that the climate change adaptation measures will have a positive effect on the increasing of the amount of herdsman's livestock and income as well as future land surface dynamics. Some suggestions on the land use management were finally proposed, which can provide significant reference information for the land use planning.

INTRODUCTION

The climate change poses great threats to the human society and natural environment [1]. In order to adapt to the climate change, social agents would take some adaptation measures that have some synergistic effects on the dynamics of land surface. However, land surface dynamics would also react on the regional climate condition. Anthropogenic adaptation measures on climate change and variability may directly or indirectly influence the dynamic land surface change [2], which will have feedback effects on climatic conditions [3–6]. One of the effects of climate change on land surface dynamics reflects on the land surface dynamics response to climate change in the Alpine region. Some human decisions can somehow mitigate the climate change by influencing CO_2 concentration according to previous studies that were conducted with the empirical or numerical methods [3, 4, 6–9], while some adaptation measures will indirectly change the climatic conditions through the intermediate impacts mainly derived from the change of land surface properties [6, 10–12]. The Intergovernmental Panel on Climate Change (IPCC) Third Assessment Report has updated a framework of interrelationship between human adaptation' design and implementation and climate change [13, 14]. However, previous studies have rarely analyzed the potential impacts of human's adaptation measures on land surface change which will result in destabilizing the climatic conditions in return [2].

There are still various difficulties in parameterizing the effects of human adaptation measures on the climate change for climate models' performance through the land surface dynamics. First, the current researches on the adaptation measures are mostly focused on the cost-benefit analysis, while there is less concern about effects of land surface change on climate which take adaptation measures into consideration [15, 16]. Second, various views and preferences should be taken into consideration when implementing the systematic simulation for land surface during the implementation process of adaptation, which involves the policies, economy, social agents, and many stakeholders [17]. Third, the research object of adaptation measures is the ecosystem that is open, complex, and full of uncertainties, the knowledge about it is still limited, and therefore there are still some uncertainties in the implementation of adaptation measures [18]. To sum up, there are still a lot of difficulties in the land surface parameterization scheme in the climate models; for example, the parameters are complex, data are difficult to capture and quantify, and so forth, while those are the indispensable contents for the study of climate models.

The case study in typical ecofragile regions is helpful to comprehensively understand the importance of adaptation measures in land surface dynamics. In this study, the Three River Source Region in Qinghai Province, which has a fragile ecoenvironment and relatively simple industry structure, was selected as the study area. The future land surface was projected on the basis of the herdsmen's adaptation measures to climate change. An agent-based model (ABM) was developed, in which the adaptation measures would be considered in land surface projection for further climate models' performance in the ecologically fragile areas.

STUDY AREA

Basic Information

The Three River Source Region, which is the headstream of Yangtze River, Yellow River, and Lancang River, is between 31 39′–36 12′N and 89 45′–102 23′E and in the southern part of Qinghai Province, China. It plays an important role as the ecological barrier for the ecological environment security and sustainable development of regions in the middle and lower reaches of Yangtze River and Yellow River in China as well as Southeast Asian countries. In the recent 30 years, the mean annual temperature is 3.96°C, with an interannual increment of 0.04°C, while the average annual precipitation is 500 mm, with a small interannual growth rate of 0.17mm. The annual average precipitation gradually decreases from east to west on a whole. In addition, the frequency and intensity of climatic disasters also tended to increase in the past decades. The frequency of droughts showed a decreasing trend after 1961, but the duration time of droughts has extended, which exert great adverse impacts on husbandry production. The temperature has increased significantly since 1998, and the winter snowfall and frequency of snow disaster have also increased. Meanwhile, there is an obvious periodical change in the storm, that is, the storm frequency in Qinghai Province has increased notably since the 1990s, especially in the eastern husbandry area. The grasslands occupy the largest area in the study area, reaching 2.665×10^5 km^2 and accounting for about 71.50% of the total land area of the study area. While the unused lands account for 16.50%, the forest lands and waters account for 9.20% and 2.10%, respectively. In the past 20 years, the cultivated lands and unused lands expanded by 46.2% and 0.64%, respectively, while the forest lands and grasslands have decreased by 0.30% and 0.28%, respectively. In 2009, the population was 7.5×10^4, with the urbanization rate reaching 15.73%. Among the 16 counties in the study area, more than half of them are key counties for poverty alleviation and development work, where the poor accounted for 63% of the total population. The animal husbandry is the main economic base for survival and development of pastoral minorities; the output value of which accounts for 57.28% of the total agricultural output.

Herdsmen's Adaptation to Climate Change

The adaptation of the farming and animal husbandry to climate change has been regarded as one of the key research spots, and adaptation measures are of great importance for mitigating the adverse effects of climate change [19]. The Three River Source Region is an important base for the development of animal husbandry in China, and it is very sensitive to climate change. The local social agents have various measures to adapt to the climate change, which were preliminarily classified mainly from the perspective of technology, engineering, management, and policy on the basis of achievements of previous researches on the adaptation measures on climate change [20]. The herdsmen's adaptation measures will have different effects, the spatiotemporal heterogeneity of which is mainly reflected by the resources endowment, population, climatic conditions, different social agents, cost-benefit, and other hidden effects that are not easy to be captured [17, 21, 22].

Herdsmen's decision on the adaptation measures can greatly affect the land surface dynamics [23]. There may be twofold impacts of these adaptation measures on the land use dynamics. Herdsmen's choice of the adaptation measures on climate change can affect their income, and the change of income will in turn affect the herdsmen's choices to adapt to the different cost, then it will further affect the decision-making behavior of the social agents. It has been discussed how economic benefits affect decision-making behavior of agents in previous works, with economic benefits represented with the cost of adaptation measures and its potential benefits in the ABM model [24, 25]. We captured the economic benefits of climate change adaptation measures with the positive mathematical programming (PMP) model. In addition, this study also involved some adaptation measures that may have impact on land surface dynamics directly, for example, the effective artificial maintenance on grassland, which can prevent the degradation from grassland to desert or even wilderness.

The adaptation measures, which are the most typical and most concerned by the herdsmen, were selected from each category based on the household survey (Table 1). It shows that the major adaptation measures in the Three River Source Region mainly include the pest control, artificial maintenance of grassland, early warning mechanism of snow disasters, and fiscal subsidies, which represent the impacts of adaptation measures from the perspective of technology, engineering, extreme weather prevention and management, and fiscal subsidies, respectively. During the survey interview, the herdsmen were asked about their preferences for adaptation measures on climate change, choice probabilities, and thinking ways, based on which the behavior rules were defined [26] and lay the foundation for the land surface parameterization scheme in the climate simulation. Additionally, the ABM model is used to simulate the effects of interaction between different social agents under particular conditions.

Table 1: Descriptions of various adaptation to climate change

Categories	Adaptation	Descriptions	Effects
Engineering measure	Artificial maintenance of grassland	Combining captive breeding with stocking.	The captive breeding is conductive to the protection of natural vegetation, while rescuing physical consumption of livestock contributes to the growth of livestock and directly increases the benefit. Meanwhile, a large-scale captive breeding has the potential to change the original pattern of land use [27].
Technical measure	Pest control	Combining biological control and chemical control.	It is helpful to control the desertification and degradation of grassland, protect grassland resources, maintain the ecological balance of grassland, and increase the income of herdsmen [28].

Management measure	The early warning mechanism of snow disasters	Early warning of monitoring of snow disaster and layout of snow equipment. Management and emergency plan of snow disaster.	Early warning work of disaster can help farmers effectively and make them take measures in time to reduce the loss of disasters. But the setting of monitoring facilities and monitoring point will increase the area of built-up land [29].
Policy measure	Fiscal subsidies	Subsidy policy of livestock breeding, subsidy of good seeds of artificial planting forage seed, policy of production subsidies, subsidy policy of animal husbandry machinery purchase, and subsidy policy of grazing prohibition for nurture.	Policy measures can avoid the herdsman's blind pursuit of profit maximization which may destroy the ecological balance. On the premise of guaranteeing the herdsman's income, the government subsidy can help the implication of policy and guide the farmer and herdsman to maintain the reasonable pattern of land use [30].

Data Collection and Processing

The household survey data used in this study were obtained from the questionnaire survey, which was implemented during August and October, 2012, in three counties of the Three River Source Region. First, three counties were selected from Tibetan Autonomous Prefecture of Golog, Tibetan Autonomous Prefecture of Huangnan, and Tibetan Autonomous Prefecture of Hainan. The selection of the sample counties was mainly based on three factors, that is, per capita grassland area, population density, and per capital GDP. Second, nine towns were selected in the three-sample counties with stratified random sampling. Then, two towns with the most and least grassland area were selected from the 9 towns, and thereafter two villages were selected in each sample town according to the employment situation in sectors except the animal husbandry. Finally, 10 herdsmen families were selected as the sample families from the herdsmen families with the large, medium, and small grassland area in each sample village. The questionnaire survey was carried out on 200 herdsman families, and the response rate of the questionnaires was 100%, with 193 valid questionnaires and a ratio of valid questionnaire of 96.50%. The contents of the questionnaire are as follows: population, land area (mainly including the grassland area), livestock production, sale of animal products, investment into the prevention and control of the damage from rats, subsidies granted for policy considerations, employment situation of sectors except animal husbandry, income from animal husbandry, and so forth.

In addition, the supplementary data consist of the basic geographic information data which are supported by Data Center for Resources and Environmental Sciences, Chinese Academy of Sciences (RESDC); the data mainly include land use data, DEM data, and topographic location data. The socioeconomic data are obtained from the Qinghai Statistical Yearbook, China Agriculture Yearbook, and China Animal Industry Yearbook. The natural environmental data and climate data are collected from the meteorological stations.

LAND SURFACE PARAMETERIZATION SCHEME

Estimation Measures of the Impacts of Herdsmen's Adaptation

It is necessary to quantitatively analyze the economic benefits of the herdsmen's decision-making behavior about adaptation measures in order to better provide reference for the land surface parameterization scheme in the ABM model. There are many methods for quantitative assessment of adaptation measures, and this study used the PMP model proposed by Howitt in 1995 [31]. This model is somehow a popular approach to analyze the policy oriented environmental and economic problems, using even scarcely available information. The method has great flexibility and conforms to the basic hypothesis of economics of diminishing marginal returns, and these advantages support wide application of PMP model in policy effect analysis [32, 33].

There are mainly three steps to analyze the cost and benefit of the adaptation measures for climate change [25]. First, the constraint conditions were set in the linear programming model according to the observed values of the baseline period, and the dual prices (shadow prices) of the constraint conditions were further calculated with the linear programming model. Second, the slope of the average cost function in the objective function of the PMP model was calculated according to the dual price of the constraint conditions. Third, the corresponding parameters in the model were adjusted according to the need of the assessment of specific adaptation measures, and the parameters were obtained and then put in the objective function to obtain the optimal solution of the nonlinear programming for the investment into adaptation measures.

Finally, in order to analyze the impacts of the adaptation measures on the income of herdsmen, the PMP model was established on the basis of the household survey data of the herdsmen families as follows. The objective function is as follows:

$$TMG = \max \sum \left(p_i y_i + sub - \left(\alpha_i + \frac{y_i}{2x_i} \right) \right) x_i. \tag{1}$$

Constraint conditions are as follows:

$$\sum x_i \times l_i \leq L, \tag{2}$$

$$\sum x_i \leq Area, \tag{3}$$

$$\sum x_i \times q_i \leq l, \tag{4}$$

$$\begin{aligned} y_i = {} & \beta_0 + \beta_1 dtpro + \beta_3 POP + \beta_4 NPP \\ & + \beta_5 dtwater + \beta_6 dtrosd + \beta_7 rain\_cv \\ & + \beta temp\_cv + \beta_9 t_0 + \beta_{10} t_{10} + \beta_{11} conrodent \\ & + \beta_{12} aprotect + \beta_{13}\, supint + \mu, \end{aligned} \tag{5}$$

where TMG is the herdsmen's family income from livestock breeding; y_i is the number of livestock per unit area of grassland;x_i is the area of grasslands for the livestock;p_i is the selling price of the livestock;α_i is the cost per unit area of grassland;y_i is the shadow price;l_i is the labor input into each livestock;q_i is the capital income per unit area of grassland; dtpro is the distance to capital of province; Gld is the area of grassland; POP is population; NPP is the net primary production; dtwater is the distance to water area; dtroad is the distance to road; rain_cv is the coefficient of variance of rain; temp_cv is the coefficient of variance of temperature;t_0 is the cumulative temperature above 0°C;t_{10} is the cumulative temperature above 10°C; conrodent is the rodent control; aprotect is the manual maintenance of grasslands; supint is the grazing intensity supervision;μ is the residual error.

In the PMP model mentioned previously, the objective function is defined as the maximization of the income of the herdsmen's family from livestock breeding for the following reasons. First, these adaptation measures implemented in the ecologically fragile areas aim to reduce the livestock loss of the herdsmen, and they consequently have positive significant impacts on the income of the herdsmen's family from livestock breeding. Second, there is a very low proportion of the income from crop production and other agricultural industries except animal husbandry in the total income of the herdsmen's family since the animal husbandry is the dominant industry in all the three counties where we do household survey in the Three River Source Region. Third, this study mainly focused on the impacts of these adaptation measures on the income of the herdsmen's family.

The income per unit area of grassland equals the product of capital income per unit area of grassland q_i, and the area of grasslands for grazing the livestock x_i, (2) expresses the constraint conditions of the labor; (3) expresses the constraint conditions of the grassland area; that is, the sum of the area of grasslands for grazing the livestock cannot exceed the total land area of the herdsmen's family; (4) is the constraint conditions of the capital; (5) provides the number of livestock per unit area of grassland under specific condition. Considering that in the system are mainly the herdsmen who graze the cattle and sheep and will not invest the income from other sources into animal husbandry, so the upper limit of total amount of investment into the livestock production should be set as the total income from livestock.

Parameterize the Adaptation in Land Surface Projection

According to different agents, the parameters of main decision-making behavior in the ABM model are different to simplify and simulate the adaptive behavior choice and the influence of each agent on climate change. Different agent has different preference for various adaptation measures, resulting in the different land surface dynamics effect. Therefore, this paper classified and extracted the agents in the study area and then determined their preference for adaptation measures according to the investigation and parameterized them.

Firstly, classify the counties and cities in Three River Source Region into three categories according to the analysis of their social and economic development, natural environment characteristics, and resource endowment characteristics; then the herdsmen's agents that are extracted from each category, mainly includes herdsmen's agent, half-herdsmen's agent, and restrictive herdsmen's agent. Secondly, design and determine the rules of agents' behavior through on-the-spot investigation about herdsmen's willingness to take climate change adaptation measures and other social economic and ecological environment consciousness, combined with the cost-benefit analysis of the adaptive measurements.

In addition, through the survey, it is found that in the Three River Source Region, the output of grasslands was relatively higher compared with that of other land use types. The grassland is also the main ecosystem in the study area, its ecosystem services function, such as livestock pasturing, water conservation, water purification, climate regulation, are indispensable to local individuals and communities.

Therefore, the relative social agents are more willing to preserve and restore the grassland.

LAND SURFACE DYNAMICS RESPONSE TO ADAPTATION FOR CLIMATE CHANGE

Economic Analysis of Adaptation Measures

This study analyzed the economic benefits of adaptation measures for climate change with PMP model and explored the impacts of different adaptation measures on the livestock production and herdsmen's income based on the different context of climate change. The results show that the livestock production and the economic benefits have a spatial heterogeneity and scale effects (Figure 1).

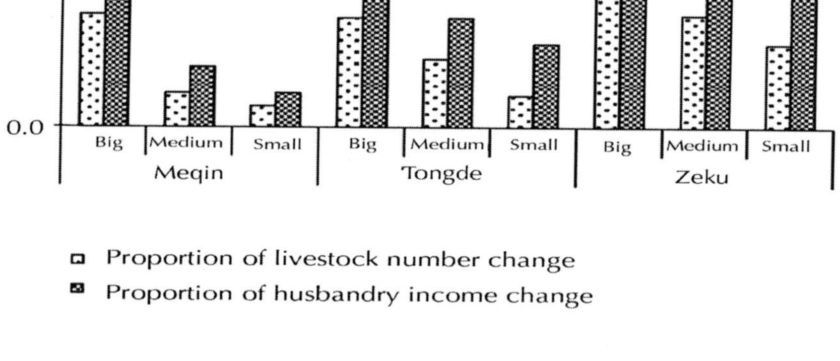

(a)

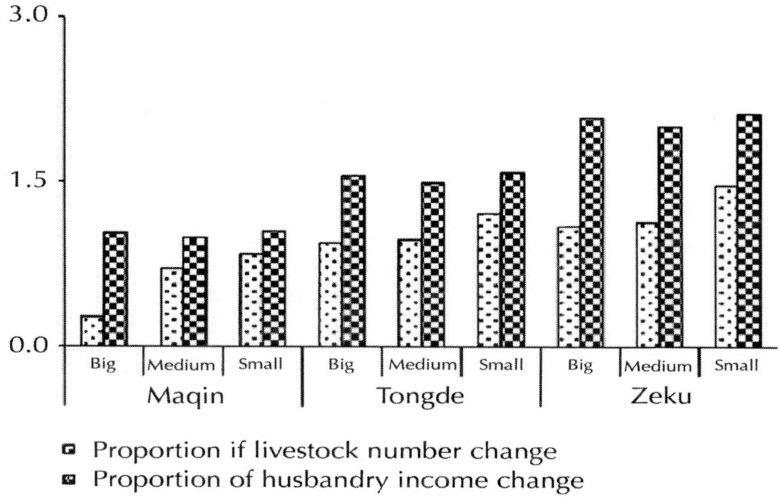
3.0
1.5
0.0
Big
Medium
Small
Big
Medium
Small
Big
Medium
Small
Maqin
Tongde
Zeku
Proportion if livestock number change
Proportion of husbandry income change

(b)

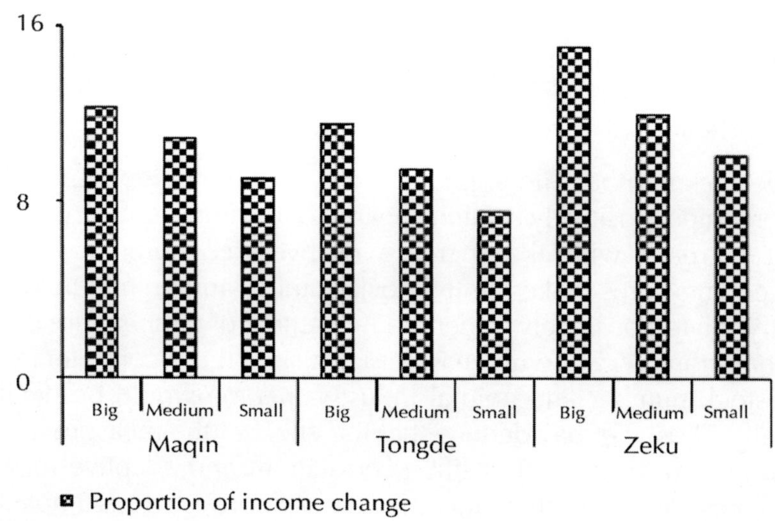
16
8
0
Big
Medium
Small
Big
Medium
Small
Big
Medium
Small
Maqin
Tongde
Zeku
Proportion of income change

(c)

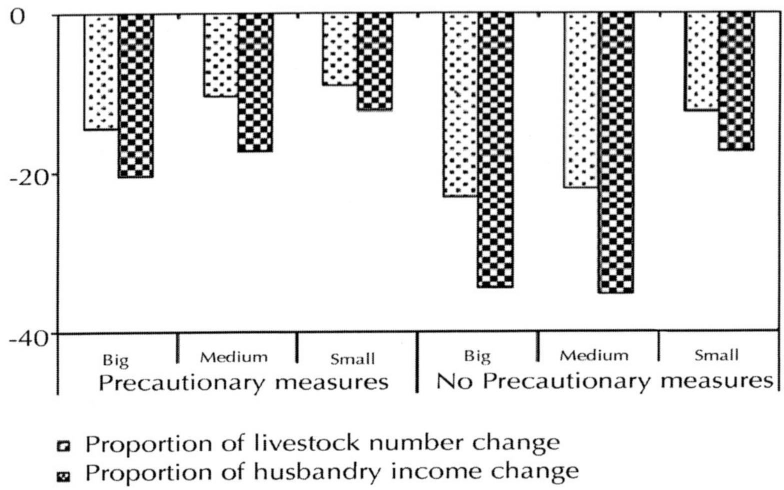

(d)

Figure 1: Mathematical analysis of herdsmen's adaptation to climate change including the technical measures (a), engineering measures (b), government subsidies (c), and extreme weather adaptation (d), from three-level pasture scales, such as large (0.67km^2–4.33km^2), medium (0.13km^2–0.67km^2), and small (0.01km^2–0.13km^2).

Livestock number and animal industry income in the same county are correlated positively, which means that animal industry income would increase with the increase of livestock number. Adaptive technical measures make the livestock number and animal income of industry related positively under the high effect of temperature change rate condition. With the decrease of farm size, the income increment of livestock number and animal industry income would be declined gradually. Thus, we may deduce that the effect of the measure and the size of farm have a good positive correlation, and adaptive measure could increase herdsmen's income, but a more obvious impact on economic benefits of large-scale pastures. The implementation of nonengineering adaptation measures plays 0.8–1% of the effect on the animal husbandry income of the herdsman after estimating based on survey data. It is also found that the impact of nonengineering adaptation measures on income of animal husbandry is different due to the pasture size. Under the high effect of precipitation change condition, increment of livestock number would increase gradually with the decrease of grassland scale in the same county. However, the variation of animal industry income would decrease firstly and then increase with the decline of grassland scale, and the income increment of small-scale grassland is the biggest. Hence, engineering measure may develop the livestock production of the herdsman and the income of animal industry. The utility of engineering measure is related positively to grassland scale, but utility effect of water resource in small-scale grassland would be bigger, thus engineering measure has a bigger impact on livestock number and animal industry income in small-scale grassland. Abnormal climate may lead to the large-scale reduction on livestock number and animal industry income; however, herdsman's income may reduce one-third of loss under the condition of early warning mechanism. At the same time, the risk management is divided into 3 levels, and we evaluated whether they develop the management of early warning at the government level to adapt to climate change using PMP model. Simulation analysis shows that the establishment of early warning mechanism can avoid about 12% of economic loss in the large-scale grassland. The loss of small-scale grassland would be the minimum compared with the other two kinds of grassland when analyzed from the scale of grassland, that means that small-scale grassland has a good flexibility, and it can adjust the scale of production timely to minimize the losses. Government subsidy has a strong positive relationship with different scale of grassland and income of animal industry in the same county.

In conclusion, the four types of adaptation measures, which start from the character of climate impact factor in different typical ecologically fragile areas and can analyze filed sample data as well as getting the results, all can develop the livestock number and income of industry in different degrees from empirical analysis of PMP model. Therefore, positive adaptive measure has a positive effect on the herdsman to adapt to climate change.

Land Surface Dynamics

According to the simulation results based on ABM model, the land surface in the Three River Source Region shows some temporal and spatial characteristics during the next 50 years. Generally, the grassland area is increasing to a certain extent with an increment of 6.04%, and the unused land gets moderate development and utilization, so the unused lands area experiences a reduction, with a rate of 14.64%. Besides, the forest land area has some decrease and other lands use type almost maintain the same (Figure 2).

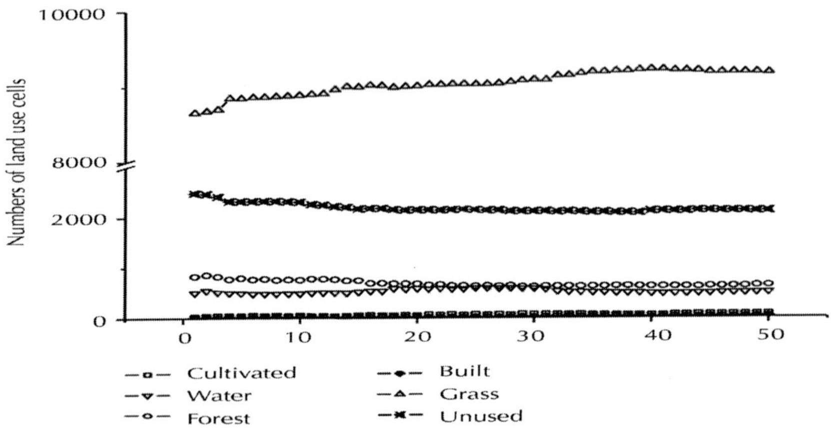

Figure 2: The underlying land surface structure in the next 50 years. The grid cell size is 2km × 2km.

In the next 50 years, the area of changed land accounted for 6.65% of the total land area, which is dominated by the conversion from unused lands and forest lands to grasslands, accounting for 2.99% and 1.81% of the total land area, respectively. Specifically, as to the transformed lands, 74.92% of forestlands and 80.6% of unused lands are converted to grasslands, in contrast, the increasing of grassland area mainly comes from the forestland and unused land, with 34.39% and 56.67%, respectively. The dramatic changes in grassland area are mainly because farms mainly pursue profit maximization and focus on considering the economic benefits of land use on the decision-making process. Grassland area would increase gradually and other land types would show a decreasing trend according to the simulation of the next 50 years. It means that the utilization way of land by farmers and herdsmen becomes simplex gradually, and dominant land use type is more and more obvious. Since single type of land will inevitably lead to the instability of ecological system, farmers and herdsmen should change the traditional way for adapting the nature, utilize adaptation measures rationally, not just pursue the revenue maximization of livestock yield and income of animal husbandry, but fully consider the rationality of land use pattern.

The distribution of land surface change simulated by ABM is showed in Figure 3. The land use change mainly occurred in the central and east part of the study area; the east part was mainly dominated by the conversions from forest lands to grasslands and from grasslands to unused lands, while the central part was manifested by the development and utilization of unused land, thus showing the conversion from unused lands to grasslands. That is because of the restively low altitude and flat terrain in the eastern area which are relatively easy for reclaiming of unused land, and there is restively high precipitation which is advantageous to the pasture production and livestock farming. These favorable factors make the herdsman develop animal husbandry by reclaiming forest land and unused land to increase the income of animal husbandry.

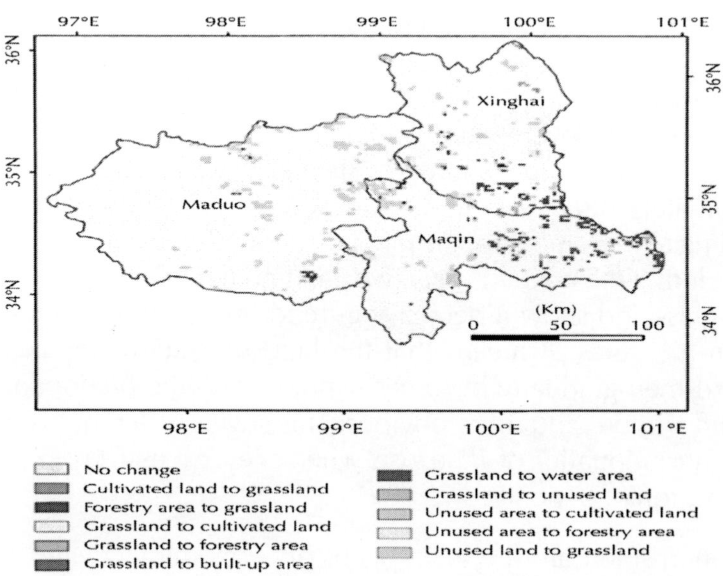

Figure 3: The spatial pattern of land surface change in the next 50 years. Only large proportion of land surface change is showed in this picture.

CONCLUSIONS AND DISCUSSIONS

This study analyzed the economic benefits of the adaptation measures on climate change with the PMP model based on the survey data in the Three River Source Region, and then projected the land use change with the ABM model. This method may provide some reference for the research on the climate simulation with different underlying land surface. The simulation result shows that the four kinds of adaptation measures will have different impacts on the development of the animal husbandry and economic benefits. Within a certain range, the larger the scale of grassland is, the more the income is. Therefore, the scale of pastures should be maintained at a certain appropriate range, and the government should continue to promote the return of the farmland to grassland and ecological migration projects. Besides, it is urgent to carry out more research on monitoring of ecological environment and ecosystem restoration, promote the comprehensive management of the ecological environment, and control the destruction of the ecological environment caused by herdsmen. According to the analysis with the ABM model on the basis of the herdsmen's adaptation measures to climate change, we could know that the adaptation measures will have synthetic impacts on the evolution of the underlying surface and promote the development and utilization of unused land and conversion to the forest lands and grasslands. The grassland increased by approximately 6.04%, while the unused land decreased by 14.64%. According to the simulation result of the land use in Three River Source Region during the next 50 years, we can find that all the land use types except grassland show a decreasing trend under the condition of adaptation measures. It means that the land utilization way of farmers and herdsmen gradually becomes simplex, and the predominance of grassland is more and more obvious. This suggests that herdsmen will ignore the rationality of land use when they excessively pursue the maximization of benefit.

The herdsmen are the main agent of economic activities and the basic unit of decision-making on the adaptation to climate change in the grazing districts, the decision-making behaviors of which have profound impacts on the land surface dynamics. The animal husbandry is the main economic source in the study area, and the adaptation measures on climate change will affect the economic income of herdsmen and of the land surface dynamics since the herdsmen's income will have effects on their decision-making on the adaptation measures and consequently lead to different land use change. This study analyzed the impacts of adaptation measures on climate change, but the land use classification is relatively rough, and the resolution is low, and there are many other factors influencing the land surface dynamics; therefore, it is necessary to implement some further researches in the future.

Although there are some uncertainties in the result, it can still provide value reference information for the land use management. Based on the result mentioned previously, some measures and suggestions of land use were put forward from all aspects. It is necessary to implement a number of measures in the overgrazing zones, for example, the reasonable land use planning, intensive development of land, control of excessive reclamation of grassland, and policy of returning grazing land to grassland, and grazing prohibition. Meanwhile, the government should provide food and economic compensation to the farmers and herdsmen so as to avoid the conversion from grassland to desert and other unfavorable land use types due to excessive degradation. Meanwhile, it is urgent to strengthen the prevention and control of pests, reduce the loss of grassland, and promote the utilization of existing grassland so as to further control the reclamation of grassland. Moreover, it is also urgent to establish the early warning mechanism of natural disasters in view of the increased frequency and intensity of natural disasters and their direct impacts on the herdsmen's economic benefit. The forest accounts for a relatively small proportion of the total area of the Three River Source Region; however, it provides abundant ecosystem services; therefore, it is necessary to make effort to manage and protect the forest in order to guarantee the sustainable development.

ACKNOWLEDGMENTS

This research was supported by the National Key Program for Developing Basic Science in China (Grant no. 2012CB955700), the National Natural Science Funds for Distinguished Young Scholar (Grant no. 71225005), and Exploratory Forefront Project for the Strategic Science Plan in IGSNRR,CAS (Grant no. 2012QY009).

REFERENCES

1. H. O. Pörtner and A. P. Farrell, "Ecology: physiology and climate change," Science, vol. 322, no. 5902, pp. 690–692, 2008.
2. J. Smithers and B. Smit, "Human adaptation to climatic variability and change," Global Environmental Change, vol. 7, no. 2, pp. 129–146, 1997.
3. M. Wang, X. Zhang, and X. Yan, "Modeling the climatic effects of urbanization in the Beijing-Tianjin-Hebei metropolitan area," Theoretical and Applied Climatology, vol. 113, no. 3-4, pp. 377–385, 2013.
4. R. A. Pielke Sr., "Land use and climate change," Science, vol. 310, no. 5754, pp. 1625–1626, 2005.
5. Y. Malhi, J. T. Roberts, R. A. Betts, T. J. Killeen, W. Li, and C. A. Nobre, "Climate change, deforestation, and the fate of the Amazon," Science, vol. 319, no. 5860, pp. 169–172, 2008.
6. J. J. Feddema, K. W. Oleson, G. B. Bonan et al., "The importance of land-cover change in simulating future climates," Science, vol. 310, no. 5754, pp. 1674–1678, 2005.
7. S. M. Sterling, A. Ducharne, and J. Polcher, "The impact of global land-cover change on the terrestrial water cycle," Nature Climate Change, vol. 3, no. 4, pp. 385–390, 2013.
8. R. M. Bright, R. Astrup, and A. H. Strømman, "Empirical models of monthly and annual albedo in managed boreal forests of interior Norway," Climatic Change, pp. 1–14, 2013.
9. J. D. Wickham, T. G. Wade, and K. H. Riitters, "Empirical analysis of the influence of forest extent on annual and seasonal surface temperatures for the continental United States," Global Ecology and Biogeography, vol. 22, no. 5, pp. 620–629, 2012.
10. J. Jin, S. Lu, S. Li, and N. L. Miller, "Impact of land use change on the local climate over the tibetan plateau," Advances in Meteorology, vol. 2010, Article ID 837480, 6 pages, 2010
11. G. B. Bonan, "Effects of land use on the climate of the United States," Climatic Change, vol. 37, no. 3, pp. 449–486, 1997.
12. L. Bounoua, R. Defries, G. J. Collatz, P. Sellers, and H. Khan, "Effects of land cover conversion on surface climate," Climatic Change, vol. 52, no. 1-2, pp. 29–64, 2002.

13. R. J. T. Klein, R. J. Nicholls, and N. Mimura, "Coastal adaptation to climate change: can the IPCC Technical Guidelines be applied?" Mitigation and Adaptation Strategies for Global Change, vol. 4, no. 3-4, pp. 239–252, 1999
14. B. Smit and O. Pilivosova, "The IPCC Third Assessment Report: Climate Change 2001, Working Group II: Impacts, adaptation and vulnerability," Adaptation to climate change in the context of sustainable development and equity, pp. 877–912, 2001,http://www.ipcc.ch/ipccreports/tar/wg2/pdf/wg2TARchap18.pdf.
15. R. S. J. Tol, S. Fankhauser, and J. B. Smith, "The scope for adaptation to climate change: what can we learn from the impact literature?" Global Environmental Change, vol. 8, no. 2, pp. 109–123, 1998.
16. K. de Bruin, R. B. Dellink, A. Ruijs et al., "Adapting to climate change in the Netherlands: an inventory of climate adaptation options and ranking of alternatives," Climatic Change, vol. 95, no. 1-2, pp. 23–45, 2009.
17. B. Smit, I. Burton, R. J. T. Klein, and J. Wandel, "An anatomy of adaptation to climate change and variability," Climatic Change, vol. 45, no. 1, pp. 223–251, 2000.
18. W. N. Adger, S. Dessai, M. Goulden et al., "Are there social limits to adaptation to climate change?"Climatic Change, vol. 93, no. 3-4, pp. 335–354, 2009.
19. X. Z. Deng, C. H. Zhao, and H. M. Yan, "Systematic modeling of impact of land use and land cover changes on regional climate: a review," Advance in Meteorology, vol. 2013, Article ID 317678, 11 pages, 2013.
20. B. Smit and M. W. Skinner, "Adaptation options in agriculture to climate change: a typology,"Mitigation and Adaptation Strategies for Global Change, vol. 7, no. 1, pp. 85–114, 2002.
21. J. D. Scheraga and A. E. Grambsch, "Risks, opportunities, and adaptation to climate change," Climate Research, vol. 10, no. 1, pp. 85–95, 1998.
22. D. B. Lobell, M. B. Burke, C. Tebaldi, M. D. Mastrandrea, W. P. Falcon, and R. L. Naylor, "Prioritizing climate change adaptation needs for food security in 2030," Science, vol. 319, no. 5863, pp. 607–610, 2008.

23. W. N. Adger, S. Huq, K. Brown, D. Conway, and M. Hulme, "Adaptation to climate change in the developing world," Progress in Development Studies, vol. 3, no. 3, pp. 179–195, 2003.

24. J. G. Polhill, L. A. Sutherland, and N. M. Gotts, "Using qualitative evidence to enhance an agent-based modelling system for studying land use change," Journal of Artificial Societies and Social Simulation, vol. 13, no. 2, 10 pages, 2010.

25. L. R. Izquierdo, N. M. Gotts, and J. G. Polhill, "Case-based reasoning, social dilemmas, and a new equilibrium concept," Journal of Artificial Societies and Social Simulation, vol. 7, no. 3, 2004.

26. P. M. Kelly and W. N. Adger, "Theory and practice in assessing vulnerability to climate change and facilitating adaptation," Climatic Change, vol. 47, no. 4, pp. 325–352, 2000.

27. S. C. Dong, C. J. Chang, and H. Y. Wang, "Ecological crisis and countermeasures of the Three Rivers' headstream region," Journal of Natural Resources, vol. 17, no. 6, pp. 713–720, 2002.

28. C. A. Gillies and R. J. Pierce, "Secondary poisoning of mammalian predators during possum and rodent control operations at Trounson Kauri Park, Northland, New Zealand," New Zealand Journal of Ecology, vol. 23, no. 2, pp. 183–192, 1999.

29. H. H. Einstein and R. Sousa, "Warning systems for natural threats," Assessment and Management of Risk For Engineered Systems and Geohazards, vol. 1, no. 1, pp. 3–20, 2007.

30. S. H. Eriksen and K. O'Brien, "Vulnerability, poverty and the need for sustainable adaptation measures,"Climate Policy, vol. 7, no. 4, pp. 337–352, 2007.

31. R. Howitt, "Positive mathematical programming," American Journal of Agricultural Economics, vol. 77, no. 2, pp. 329–342, 1995.

32. R. E. Howitt, "PMP based production models-development and integration," in Proceedings of the European Association of Agricultural Economists International Congress, Copenhagen, Denmark, 2005.

33. T. Heckelei, W. Britz, and Y. Zhang, "Positive mathematical programming approaches—recent developments in literature and applied modelling," Bio-Based and Applied Economics, vol. 1, no. 1, pp. 109–124, 2012.

Chapter 6

Agricultural Modernization, Structural Change and Pro-Poor Growth: Policy Options for the Democratic Republic of Congo

Christian S. Otchia

School of International Development, Nagoya University, Furo-cho, Chikusa-ku, Nagoya 464-8601, Japan

ABSTRACT

This paper applies the framework for pro-poor analysis to welfare changes from a CGE-microsimulation model to analyze what are the better or worse models for agriculture modernization, and to estimate the contribution of growth and redistribution to changes in poverty in DRC. The findings indicate that labor-using technological change generates absolute and relative pro-poor effects whereas capital-using technological change leads to immiserizing growth. More importantly, the results suggest that labor-using technological change can be independently sufficient for reducing poverty via the income growth effects. This study also highlights how developing input supply networks, securing tenure among smallholders, and improving access to land for women are important for pro-poor agricultural modernization.

INTRODUCTION

Agricultural transformation is essential for the Democratic Republic of Congo (DRC) because it has huge potential to spur growth and raise income. Agriculture employs most of the labor in DRC and produces the largest percentage of total value added. Figure 1 shows that agriculture employs 60.2 percent of the Congolese labor force and generates about 21 percent of total value added. Sectors such as textiles, chemicals, construction, and forestry only produce a small share of value added and contribute marginally to employment creation. Figure 1 further indicates that agriculture and trade sectors lie below the 45-degree line, meaning that the share of employment in these sectors is higher than the share of value added from these sectors. However, the largest gap between the contribution to value added and employment appears to be in agriculture. This indicates that agriculture has the lowest productivity in DRC's economy.

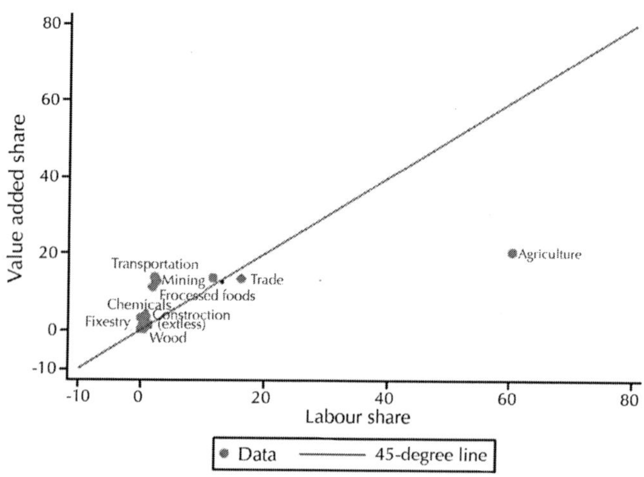

Figure 1: Profile of sectoral employment and value added (2005). *Source*: Author's based on DRC national accounts (2005).

Agriculture is the most unproductive sector in DRC because of inconsistent and uncoordinated agricultural development strategies, coupled with conflict and the progressive withdrawal of the government from supporting agricultural activities. According to Otchia ([2013b]), government policy implemented since 1966 led to the collapse of large-scale commercial agriculture, favored subsistence agriculture, and distorted economic incentives against agriculture. In addition to this, the government removed all subsidies and price support measures to agriculture in 2002. Consequently, farmers use a rudimentary agricultural technology mostly based on outdated production methods and inputs. Agriculture also faces high transaction costs due to the lack of infrastructure most of which was destroyed during political conflicts. Low productivity in agriculture entails unstable and low paid jobs. As a result, an overwhelming proportion of agricultural workers are poor. Four out of every five rural poor work in agriculture. In urban areas, agriculture accounts for one-third of the poor.

Nevertheless, agriculture is still attracting labor in both urban and rural areas. According to Herderschee et al. ([2012]), agriculture provided employment for 10 million people in 2005 and 15 million in 2010. Despite the low productivity, labor accrues in agriculture because it can produce the amount of food necessary for their subsistence. This implies that most of the farming activities are of a small scale and aim to increase food security. Given its low productivity, increasing the amount of labor and land is the only way to raise production in agriculture. Labor flows to subsistence farming as it uses essentially manual work, whereas large-scale farmers tend to expand land. Indeed, DRC is far from reaching the agriculture frontier, as it uses only 11 percent of the 80 million hectares of arable non-forest land for agriculture. However, in recent years, much of agricultural land has been developed for export-oriented large-scale commercial agriculture. These agricultural investments are made by foreign investors to secure their own food needs. This constrains access to land for small-scale farmers.

Against this background, agricultural productivity improvement is the fundamental policy to initiate agricultural transformation and raise income of the poor (Alvarez-Cuadrado and Poschke [2011]; Ngai and Pissarides [2007]). The reason is that productivity improvement "pushes" labor out of agriculture and increases farmers' real wages; "pulls" jobs in sectors that use agriculture as inputs; and increase supply of affordable food in the economy. The empirical literature reports strong and robust effects of agriculture productivity on poverty (Thirtle et al. [2003]; Irz et al. [2001]; de Janvry and Sadoulet[2010]). However, the magnitude of poverty reduction due to agricultural productivity growth varies largely across countries, depending on the way they developed and used new technologies (de Janvry and Sadoulet [2010]).

The literature documents a range of policies to increase agriculture productivity and enhance income-increasing structural change. Among them, technological change has been acknowledged as the principal driver of productivity growth (OECD [2012]; Morris et al. [2007]; DFID [2006]). However, it is worth mentioning that the innovation, selection, and adoption of new technologies depend on the agriculture frontier, factor endowment, and market imperfections. Hayami and Ruttan ([1970]) used data on agriculture inputs to assess how endowment drove the direction of technical change in the US and Japan during 1880–1960. They found that land abundance in the US favored labor-saving technological change while the land scarcity in Japan led to the development and adoption of land-saving technologies. As a mechanization strategy, labor-saving technological change consists of using tractors and machinery, whereas land-saving technological change focuses on biological and chemical innovations. A recent successful case of land-saving technological change occurred during the Green Revolution in Asia. The Green Revolution was an intensifying of input-based production characterized by the use of high-yielding and fertilizer-efficient new varieties of seed (rice and wheat). Policymakers initiated this type of agricultural transformation to increase food production and reduce hunger and malnutrition in the 1960s. Hence, it is conceptually clear that the Green Revolution increased agriculture and food production. Empirical results also indicate that it led to poverty reduction as it raised farmers' income and increased food affordability.

Though it is expected that agricultural productivity improvement tends to reduce poverty, the extent to which it reduces inequality and benefits small-scale farmers is still open to question. For instance, the pro-poorness of the Green Revolution has been disputed, since its effectiveness in reducing inequality is not straightforward. The main argument states that the Green Revolution worsened income distribution as it was biased in favor of larger farmers and missed the poorer subsistence small-scale farmers (Das [1998]; Griffin [1979]; Freebairn [1995]; Goldman and Smith [1995]). Furthermore, it increased landless farmers and the demand for unskilled labor, which in turn lowered wage laborers (Hazell and Ramasamy [1991]; Glaeser [1987]; Cleaver [1972]).

Despite this, the experience of Asia points to a clear consensus on the role of strong public policies and investment in creating a pro-poor Green Revolution (Eicher [1995]; Smale [1995]; Hazell[2009]). These policies include agricultural research and development, irrigation, rural roads, access to credit, and price support policies. In addition, those policies had been successful when they have been implemented together. However, there is no empirical assessment on the pro-poorness of technological change and the complementary rural development policies in Africa, especially in DRC.

This paper thus aims to assess what are the better and worse models for agricultural modernization in DRC. Agricultural transformation is qualified as a better model only if it is centered on small-scale farmers as most of them are poor and have limited resource endowment relative to other farmers. To put it differently, a better model for agricultural modernization produces pro-poor effects where poor households gain relative to the richer ones. Several recent studies have looked at the pro-poor effects of policies, particularly using CGE-microsimulation model (Boccanfuso et al., [2011], [2013a],[2013b]; Annabi et al. [2008]; Ravallion and Lokshin [2008]). Most of the studies do not show factors behind the differences in the impacts of policy on pro-poor growth or decompose the changes in poverty into growth and distribution components, but rather show how poor benefit/lose relative to rich segments of the population. Boccanfuso and Kaboré ([2004]), however, did find that the relationship between poverty, growth, and inequality relationship is heterogeneous and conditional on context.

To look at the pro-poorness of different strategies for modernizing agriculture, I combine three techniques, namely a computable general equilibrium model, a household-survey based microsimulation, and least square regressions. I adopt a sequential approach that can be described in four steps. In the first step, I evaluate the effects of agricultural modernization strategies on employment, wages, and rents, and the price of goods and services. I use a CGE-microsimulation model that captures various links through which agricultural modernization affects households. These links include the return to labor and land, the price of goods, the impact on non-agriculture sector, and sectoral labor mobility. Then I feed the changes from the CGE model into a microsimulation model, which takes into account household heterogeneity in terms of factor endowments and consumption patterns, to generate welfare gains or losses at the household level. Using these welfare changes, in the third step I apply the pro-poor growth framework to assess which of the agricultural modernization strategies is pro-poor and the extent to which growth and redistribution contribute to welfare changes, following Annabi et al. ([2008]). Finally, I select a strategy that produced pro-poor welfare gains in the previous stage, and use a least square regression as in Ravallion and Lokshin ([2008]) to quantify the determinants of pro-poor agricultural modernization at the household level.

The rest of the paper is organized as follows. Section 2 presents an overview of the agricultural sector in DRC. Section 3 presents the theoretical framework of agricultural modernization, while Sect. 4 explains the features of the CGE-microsimulation model and presents an analytical framework for pro-poor analysis. Section 5 discusses and presents the results of policy experiments. Finally, Sect. 6 provides a summary of the results and lessons for policymakers.

OVERVIEW OF THE CONGOLESE AGRICULTURAL SECTOR

The Congolese economy depends on the agricultural sector, which contributes more than 20 percent of the country's GDP. However, it is important to note that the importance of agriculture is not a result of improved agricultural production. Rather, it is due to the marked reduction of mining production, which declined faster than agriculture. In recent years, agriculture became an urban phenomenon, especially for food security reasons and proximity to markets. Urban or peri-urban farming in the DRC is not only a response to the rise in food insecurity; it also serves as an income-generating activity because of the increasing demand for vegetables in cities and soaring food prices. As a result, the agricultural sector has become the second largest employer for urban workers after the trade sector. This section describes some key characteristics and features of agriculture in DRC, relevant to the problems under review. These are (a) land size and distribution; (b) fertilizer use; (c) production and productivity; (d) agricultural trade patterns; and (e) agriculture's contribution to poverty.

Land Size and Distribution

Land is a very important asset for DRC farmers for its economic, cultural and spiritual significance. Due to bad governance (corrupted judiciary system, weaken traditional land rights, flawed land law (uncertain land rights, outdated land registry), however, land has become the key driver of conflict in the eastern part of the country (Vlassenroot and Huggins [2005]; Huggins [2010]). The most core issue in conflicts over land concerns limited access to land, land succession problem, and inequitable distribution. There are other factors behind land issues in DRC, such as colonization, land grab, migration, and climate change (Long [2011]; Chausse et al. [2012]; African Union et al. [2012]). The consequences of these measures and events are visible in all their extent: increased landless and reduced average land size.

For instance, the highly skewed nature of land distribution in DRC is evident if one looks at Fig. 2where I plot the value of land per household, per capita and per adult across three locations, namely urban, peri-urban, and rural areas. The figures indicate that farms are very small; the average land holding per household is in order of 1.3 hectare (ha) in urban areas and around 2 ha in peri-urban and rural areas, whereas the median of land per household is 0.8 in urban areas and 1 ha in peri-urban and rural areas. The median are about 50 percent lower than the mean, implying the existence of high land inequality. Moving to the per capita distribution, panel (b) of Fig. 2 shows that average land per capita is 0.3 hectares in urban areas, while it is 0.4 and 0.6 hectares in peri-urban and rural areas. Despite the dominance of small farms, it is interesting to note that the average land per capita is not much of issue as it ranks DRC among countries with more than an average of potential agricultural land. On average, land per adult is a bit more than half a hectare in urban areas but nearly 1 hectare in peri-urban and 1.15 hectares in rural areas. As one would expect, the average land per adult is significantly higher in rural areas because of migration to urban areas.

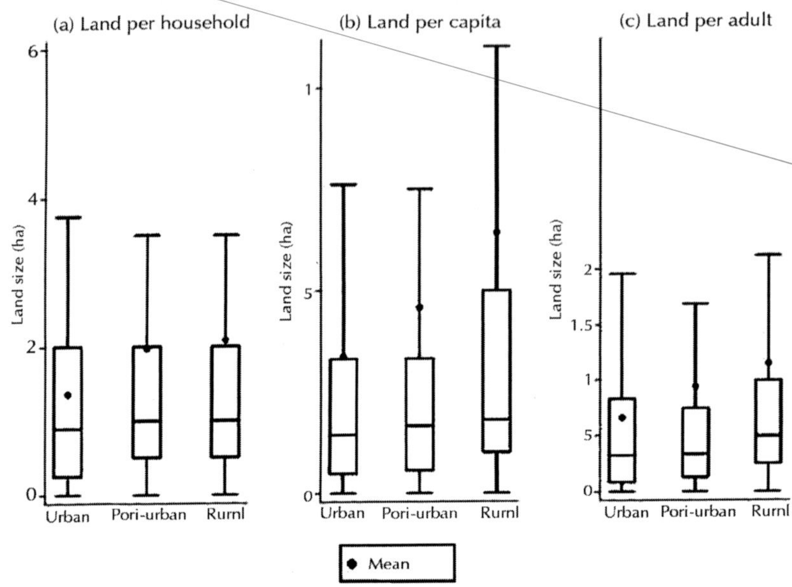

Figure 2: Boxplots for land size.

The significant discrepancies between mean and median land size suggest the limitation of the figures to assess land distribution in DRC. Therefore, I complement the land distribution analysis by decomposing the Gini coefficient of inequality between urban, peri-urban, and rural areas. In this study, I decompose the Gini coefficient into three components, namely a within-group inequality term, a between-group inequality term, and an overlap term. The within-group inequality term is a weighted sum of the inequalities calculated for each area (urban, peri-urban, rural), whereas weights depend on the population and land share of each area. The between-group inequality term is calculated on the total population where the land size of each person in the area is replaced by the average land size in the area where he lives. This component of inequality thus indicates the mean difference across areas. The overlap term is a residual term that arises because the areas' land size ranges overlap. It reflects the interaction effect among groups.

Based on the figures on Table 1, it appears that the overall Gini coefficient of land per household is 0.46, indicating that land inequality is very high in DRC. Table 1 also shows a more unequal land distribution in terms of land per capita, as the Gini of 0.56 indicates.

Comparing these estimates to those of the sub-region reported by Jayne et al. ([2003]), it appears that DRC has an unequal land distribution than Zambia and Mozambique, where the Gini index of land per household is 0.44 and 0.45, respectively, and the Gini of land per capita is 0.50 and 0.51. Jayne et al. ([2003]) report higher Gini of land per household for Rwanda (0.52), Kenya (0.55) and Ethiopia (0.55) than DRC. In addition, the distribution of land per capita in DRC is similar to Ethiopia (0.56) but more unequal than in Rwanda (0.54) or Ethiopia (0.55).

Table 1: Land distribution in DRC

Areas	Land per household			Land per capita			Land per adult		
	Gini index	Absolute contribution	Relative contribution	Gini index	Absolute contribution	Relative contribution	Gini index	Absolute contribution	Relative contribution
Urban	0.459	0.003	0.007	0.548	0.004	0.007	0.459	0.003	0.007
	(0.011)	(0.000)	(0.001)	(0.027)	(0.001)	(0.001)	(0.012)	(0.000)	(0.001)
Peri-urban	0.453	0.007	0.015	0.539	0.008	0.014	0.453	0.008	0.016
	(0.012)	(0.001)	(0.002)	(0.024)	(0.001)	(0.002)	(0.012)	(0.001)	(0.002)
Rural	0.467	0.285	0.613	0.564	0.359	0.640	0.463	0.288	0.622
	(0.005)	(0.006)	(0.012)	(0.008)	(0.009)	(0.014)	(0.005)	(0.007)	(0.013)
Within	–	0.296	0.635	–	0.370	0.660	–	0.298	0.645
Between	–	0.027	0.059	–	0.030	0.053	–	0.026	0.055
Overlap	–	0.143	0.307	–	0.161	0.287	–	0.139	0.300
National	0.466	0.466	1.000	0.561	0.561	1.000	0.463	0.463	1.000
	(0.004)	(0.004)	(0.000)	(0.007)	(0.007)	(0.000)	(0.004)	(0.004)	(0.000)

Source: Author's calculations, based on the 2005 household and informal producer survey (*Enquête* 1-2-3), Institut National de la Statistique, DRC

Otchia

Otchia *Journal of Economic Structures* 2014 3:8, doi:10.1186/s40008-014-0008-x

Furthermore, Table 1 indicates that land distribution is more unequal in rural area, as the Gini of rural area is higher for land per household, land per capital, or land per adult. Similarly, rural area is the most responsible of land inequality, as it contributes to 61 percent of total land inequality. This leads the within-area inequality become high in explaining land inequality than the between-area inequality. The high share of within-sector term calls for attention in reducing land inequality in rural sector.

Fertilizer Use

Now, I turn to the use of fertilizer in DRC. Figure 3 compares the use of fertilizers in DRC and some African countries. One can see that DRC uses less fertilizer than its neighboring countries. Between 2006 and 2010, the average intensity of fertilizer use in DRC was only 0.47 kg/ha, while it reached 46.51 and 36.69 kg/ha in South Africa and Morocco. High cost of fertilizers is the main reason that limits the fertilizer use in DRC. Most of these costs are due to imports and transportation costs, as DRC imports about 10,000 metric tons of fertilizer annually. According to Nweke et al. ([2000]), most of farmers in DRC have low incentive to invest in fertilizer because imported wheat and rice are available at competitive price in nearby commercial markets. Unavailability of credit and support price measures for dealers and farmers plays a major role in limited use of fertilizer. In fact, fertilizer import business in DRC is too small and unstable to ensure its survival.

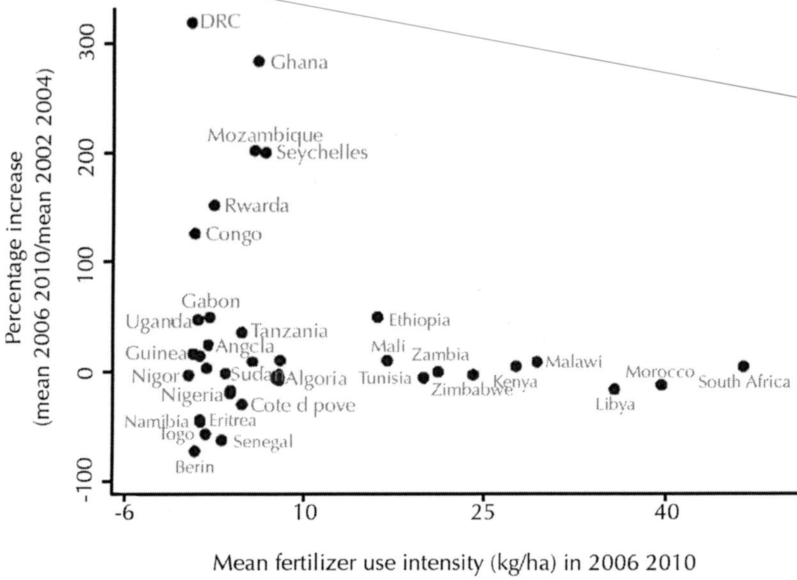

Figure 3: Fertilizer use. *Source*: Author's creation based on FAOSTAT.

The other factors for low fertilizer use are the lack of adequate knowledge about fertilizers, bad quality of available fertilizers, poor extension services, and local farming practice. Mumvwela ([2004]) stated that farmers in western DRC use also less of livestock manure that are available. Despite the low intensity of fertilizer use, it is interesting to see that DRC is rapidly increasing the amount of fertilizer. Figure 3 also shows that DRC increased by 300 percent the use of fertilizer between 2006–2010 and 2002–2004. Nevertheless, there is still much to do, as yields have not responded yet to the increase of fertilizers.

Agricultural Production and Productivity Trends

Table 2 shows the growth rates of production of the main agricultural products in the DRC between 1960 and 2010. The main food crops (cassava, plantains, and maize) accounted for 80 percent of total agricultural production, while cash crops represented less than 15 percent.

Table 2: Agricultural growth rates and area harvested (percent)

	Percent of agriculture production in 2011	Growth rates of agriculture production					Growth rates of area harvested				
		1960–1970	1970–1980	1980–1990	1990–2000	2000–2010	1960–1970	1970–1980	1980–1990	1990–2000	2000–2010
Food crops											
Cassava	67.48	17.51	26.14	14.07	−18.72	1.20	−0.77	0.52	1.10	0.24	−1.90
Plantains	6.73	22.82	23.48	−4.75	−41.13	14.35	−35.47	−58.59	−4.96	14.77	3.27
Maize	5.83	21.18	35.81	31.97	4.37	7.29	3.70	13.54	21.70	19.64	4.38
Bananas	1.77	17.58	20.20	−6.53	−15.21	13.42	−0.70	−7.48	−22.62	3.18	10.70
Rice, paddy	1.53	53.24	31.36	21.85	−18.97	4.93	5.71	−3.07	−20.57	−18.60	11.66
Sweet potatoes	1.13	11.12	19.63	2.20	−60.37	8.95	37.11	37.15	37.43	23.44	23.29
Soybeans	0.09	48.52	53.34	45.64	8.89	25.62	42.87	7.54	10.05	0.03	1.95
Cash crops											
Palm oil	0.81	−10.88	−17.74	8.81	−1.43	6.21					
Wheat	0.04	−6.55	39.67	49.62	−8.42	9.17	−8.59	−8.25	−12.56	−34.76	6.10

Source: Author's calculations, based on the FAOSTAT database of the Food and Agriculture Organization of the United Nations (FAO)

Otchia

Otchia *Journal of Economic Structures* 2014 3:8, doi:10.1186/s40008-014-0008-x

Data in Table 2 reveal a widely varying pattern of production growth rates among the different agricultural products over 1960–2010. This is the result of uncoordinated agricultural development strategies, coupled with conflict and the progressive withdrawal of the government from supporting agricultural activities. Cash crops were the backbone of DRC agriculture in the 1960s. In particular, palm oil generated half of total export earnings and made the DRC the second largest exporter of this crop in the world. As a result of a succession of policy strategies and measures, however, the production of cash crops (rubber, sugar, coffee, and cotton, in addition to palm oil) declined starting in the early 1970s. For instance, the production of palm oil fell from 224,000 metric tons in 1961 to 187,000 metric tons in 2011. This coincided with the implementation of goal no. 80 of a 10-year plan of industrialization through domestic and external loans. The collapse of cash crop production was accelerated by "Zaïrianization" (1973–1974), a policy of expropriation of foreign-owned production units by the government, which then handed them over to nationals. This policy led to the collapse of large-scale commercial agriculture, favored subsistence agriculture, distorted economic incentives against agriculture (Otchia [2013a]), and led to conflicts. Growth in palm oil production resumed in the 1990s as a result of another agricultural and rural development plan, *Le Plan Directeur*, but could not be sustained because of looting (1991–1993) and war (1998–2002).

War and civil conflict in the 1990s negatively affected production of food crops as well. Table 2indicates that sweet potatoes, plantains, rice, cassava, and bananas experienced a large drop during 1990–2000. In spite of this decline, the agricultural sector has continued to serve as the backbone of the Congolese economy. Growth of agricultural production, especially food crops, resumed during 2000–2010. Production of soybeans, which are grown extensively for their nutritional qualities, grew by 25.6 percent, while that of plantains and bananas grew by 14.4 and 13.4 percent, respectively. However, as long as production technology remains rudimentary and producers lack improved varieties and inputs, the growth of food crop production continues to depend on available quantities of the basic production factors of land and labor. For example, the harvested area of sweet potatoes and paddy rice grew by 23.3 and 11.7 percent, respectively, from 2000 to 2010.

Concerning agricultural productivity, panel (a) of Fig. 4 displays agricultural land productivity and the per capita capital stock in land development, while panel (b) plots agricultural labor productivity and per capita capital stock in machinery and equipment. As can be seen, land productivity increased between 1980–1989 and 1990–1999, but dropped afterward due to the collapse of infrastructure and frequent displacement of farmers during the war period. Labor productivity, on the other hand, decreased continuously starting in 1980–1989, and then fell drastically between 1990–1999 and 2000–2007. Land and labor productivities in DRC are low and decreasing for various reasons, including the lack of investment in accumulating capabilities, low fertilizer use, land size, war and displacement, the informal character of agriculture, and the rudimentary nature of technology used in this sector. For instance, Fig. 4 indicates that per capita stock in land, and machinery and equipment development is decreasing since 1980–1989.

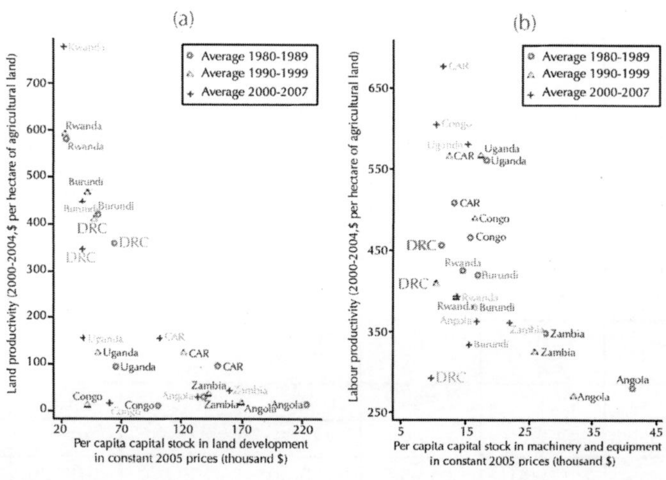

Figure 4: Labor and land productivity in DRC and neighboring countries. *Source*: Author's creation based on FAOSTAT.

Agricultural Trade Patterns

Exports from the DRC, after having more than doubled from 1961–1980, decreased sharply during 1980–2000, as shown in Table 3.

The reason is that the development policies implemented during the latter period, such as Zaïrianization, undermined the viability of large-scale agricultural projects and disrupted the maintenance of rural infrastructure and support services, as discussed in the previous section. Exports of palm oil, rubber, and cotton collapsed in the 1990s, and in later years DRC agricultural exports came to be dominated by bran of wheat and coffee, which amounted to 62.8 percent of such exports in 2010.

Table 3: Agricultural exports, selected years

	1961	1970	1980	1990	2000	2010
Total agricultural exports (thousand constant $)	107,340	112,196	234,839	139,080	39,308	75,120
Export share in total agricultural exports (percent)						
Bran of wheat	0.00	0.00	16.25	14.63	41.12	52.09
Cocoa beans	1.23	1.43	2.26	3.28	4.84	2.09
Coffee, green	8.27	16.74	40.16	63.54	50.02	10.75
Cotton lint	3.71	2.48	1.19	0.00	0.34	0.00
Palm oil	37.62	34.08	5.42	0.00	0.00	0.90
Rubber nat dry	9.18	9.86	10.55	4.25	0.27	1.03
Tea	1.23	1.73	0.79	1.59	0.15	0.16
Other	38.76	33.69	23.37	12.71	3.25	32.98
Total	100.00	100.00	100.00	100.00	100.00	100.00

Source: Author's calculations, based on FAOSTAT

Otchia

Otchia *Journal of Economic Structures* 2014 **3**:8, doi:10.1186/s40008-014-0008-x

Since the level of food production is low, the DRC dependency on imported food has increased. Table 4 indicates that food imports increased approximately 40-fold between 1960 and 2010, from $23 million in 1960 to $977 million in 2010. Major imports included flours of wheat and maize, sugar, palm oil, and meat. As can also be seen in the table, the DRC only started to import significant amounts of maize, sugar, and palm oil in 2000.

Table 4: Food imports, selected years

	1961	1970	1980	1990	2000	2010
Total food imports (thousand constant $)	22,792	61,887	156,900	241,393	214,424	977,293
Import share in total agricultural imports (percent)						
Wheat	0.03	0.00	39.60	29.36	21.34	28.66
Flour of maize	0.00	0.00	0.00	0.00	7.47	13.18
Flour of wheat	32.47	27.82	1.76	11.37	18.03	9.68
Sugar raw centrifugal	0.00	0.00	0.00	0.00	7.79	8.95
Palm oil	0.00	0.00	0.00	0.00	0.94	5.61
Chicken meat	0.65	0.41	0.95	4.30	2.52	4.19
Rice—total (rice milled equivalent)	14.81	8.52	6.19	16.60	10.95	3.41
Malt	15.43	12.43	4.76	4.88	3.14	3.31
Other	36.60	50.82	46.74	33.49	27.83	23.01
Total	100.00	100.00	100.00	100.00	100.00	100.00

Source: Author's calculations, based on FAOSTAT

Otchia

Otchia *Journal of Economic Structures* 2014 3:8, doi: 10.1186/s40008-014-0008-x

Agriculture and Poverty

Figure 5 plots the breakdown of the poverty headcount by sectors of activity, i.e. agriculture and other sectors, and compares it across urban and rural areas. This allows for evaluating the contribution of agriculture to poverty reduction. The figure clearly shows that the agricultural sector is home to the poor. In rural areas, where the poverty rate is extremely high, 83.4 percent of poor households work in agriculture, while only 16.6 percent of the rural poor work in other sectors. In urban areas, the agricultural sector accounts for 34.4 percent of the poor population, which is still very high compared to the trade and transportation sectors. At the national level, the share of poor households that work in agriculture is 63.4 percent. It can thus be concluded that high poverty rates and the recent rise in rural poverty are at least partly related to the fall in labor and land productivity in agriculture described in Sect. 2.3.

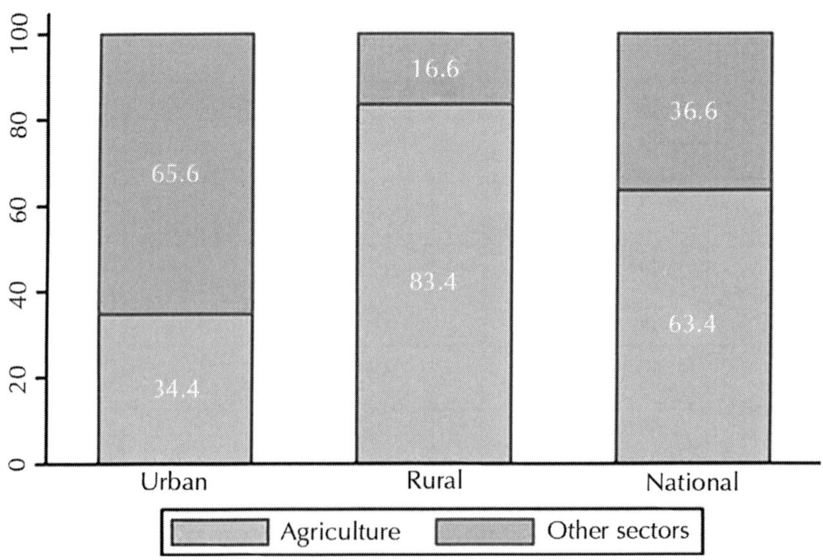

Figure 5: Distribution of poverty by economic sector in 2005 (percent). *Source*: Author's computations, based on the 2005 household and informal producer survey (*Enquête* 1-2-3).

Turning to the structure of budget shares and their distribution across groups, Table 5 reports a product-disaggregated breakdown of consumption expenditure by deciles of the income distribution. Here the interest is in examining how the expenditure allocation across different consumption items evolves with the income level of the household. Several points are worth noting. Looking first at food expenditure, it is important to highlight the importance of food consumption in Congolese households' expenditure. Table 5 shows that Congolese households allocate the highest share of their expenditure to food consumption, and that this share decreases for rich households, following Engel's Law. Apart from food consumption, the category that includes housing, electricity, gas, and water represents the second largest expenditure item. The share of this category is almost homogeneous across all households, averaging 14 percent of total expenditure. The expenditure breakdown implies that after households cover their needs in food and housing, they have little money left for other services such as education and medical care. This is especially true for poor households: as can be seen, the share of education in total expenditure for the three lowest deciles is close to 1 percent.

Table 5: Distribution of consumption by expenditure group (percent)

Products	Expenditure group (decile)									
	1	2	3	4	5	6	7	8	9	10
Food consumption	72.1	72.0	71.8	71.1	70.2	68.9	71.3	68.0	64.8	52.7
Marketed	40.1	40.2	41.8	42.3	43.9	47.1	51.7	51.5	54.8	48.5
Home-produced consumption	32.0	31.8	30.0	28.8	26.3	21.8	19.6	16.5	10.0	4.2
Beverage and tobacco	2.7	2.4	2.3	2.2	2.3	2.1	2.8	2.2	2.8	1.9
Clothing and footwear	3.5	3.9	4.8	4.8	4.7	4.6	4.2	4.7	4.3	6.0
Housing, electricity, gas, water	15.8	14.3	13.6	12.9	14.0	13.7	12.3	13.2	13.8	16.2
Medical care	2.8	2.7	2.6	2.6	2.6	3.1	2.5	3.0	2.9	4.2
Transportation and communications	0.2	0.7	0.8	1.3	1.5	2.1	1.7	2.8	3.9	8.6

Education	0.8	1.1	1.1	1.3	1.3	1.7	1.5	1.6	1.7	3.0
Recreation and culture	0.2	0.4	0.4	0.4	0.5	0.6	0.5	0.7	0.8	1.2
Restaurant and hotels	0.9	0.7	0.5	1.1	0.8	1.0	0.8	1.3	2.2	2.8
Other services	1.1	1.9	1.9	2.2	2.1	2.3	2.2	2.4	2.8	3.3
Total	100.0	100.0	100.0	100.0	100.0	100.0	100.0	100.0	100.0	100.0

Source: Author's calculations, based on the 2005 household and informal producer survey (*Enquête* 1-2-3)

Otchia

Otchia *Journal of Economic Structures* 2014 3:8, doi: 10.1186/s40008-014-0008-x

In order to obtain more detailed information on food consumption patterns, Table 5 disaggregates food expenditure into market goods consumption and home-produced consumption. The food consumption pattern varies significantly using this disaggregation. Market food consumption represents 40 percent of poor households' expenditure, and this share increases with income. This means that rich households spend a larger share of their income on market goods than poor households. Looking at the home-produced consumption pattern also provides some important insights for policymakers. Home-produced consumption represents 32 percent of total expenditure for the poorest decile, which is approximately half of their food consumption expenditure, but this share declines significantly with income. It is 26.3 percent for the fifth income group decile, 10 percent for the ninth decile, and 4 percent for the richest decile.

Figure 6 extends the expenditure analysis by plotting the kernel density estimates of urban and rural households for food consumption. The figure plots the estimated density function of food consumption per adult equivalent for urban and rural households. It can be clearly seen that the distribution of the log of food consumption per adult equivalent for urban households is skewed to the left, while for rural households, the distribution is slightly skewed to the right. Two vertical lines represent the food expenditure poverty line for urban and rural areas. This enables one to assess the potential impact of growth on poverty reduction. The figure shows that the distance between the poverty line and the mode of urban per capita expenditure distribution is not large. From a poverty reduction policy perspective, this implies that it would require only a very small increase in per adult equivalent food consumption to move many households out of poverty in urban areas. In rural areas, however, the mode of the density function is quite far from the rural food poverty line. This indicates the need for poverty reduction policies capable of increasing incomes of the poor more in the rural sector than in the urban sector in order to achieve similar reductions in poverty rates.

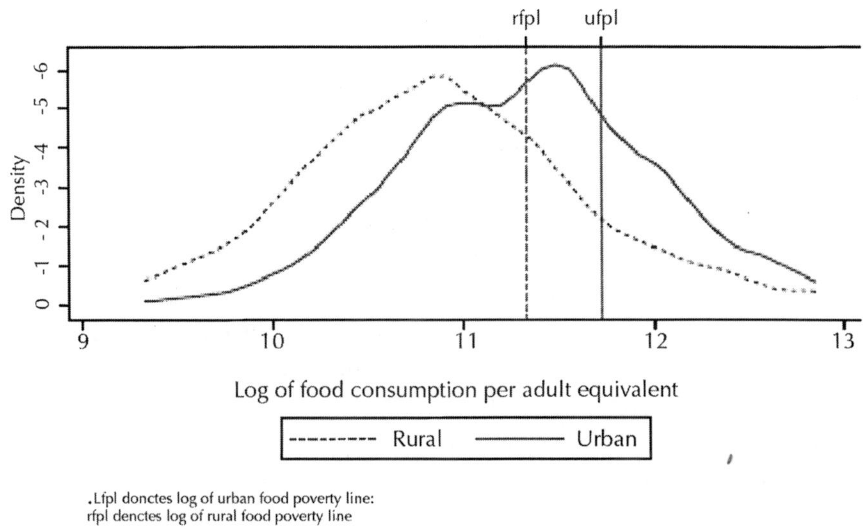

Figure 6: Kernel density function of households by per adult equivalent food consumption *Source*: Author's computations, based on the 2005 household and informal producer survey (*Enquête* 1-2-3).

THEORETICAL CONSIDERATION

The literature on technological change in agriculture documents that countries tend to adopt the technology that can raise the productivity of the scarce factor or the factor with the lowest quality. Countries with scarce labor but abundant land and capital tend to adopt labor-saving technologies such as tractors and machinery. Countries with scarce land but abundant labor tend to adopt land-saving types of technologies such as chemical and biological high-yield technologies. Hayami and Ruttan ([1985]) provide a theoretical framework of this type of biased technological change based on the induced innovation hypothesis. According to this theory, innovation is induced as a response to changes in relative prices, which push firms to innovate in order to use less of the resource that has become more expensive. However, the hypothesis of biased technical change as hypothesized by Hayami and Ruttan may not hold in low income and sub-Saharan countries (Cuffaro [1997]). In most of these countries, land and labor are abundant, but capital is scarce, and land inequality is high so that most of the farmers are smallholders. Thus, the theory of induced innovation cannot hold for the following reasons: (1) demand for innovation for small- and large-scale farms is different; (2) small- and large-scale farms have different influence on public research; (3) imported technology is absent in induced innovation theory.

In this study, the concept of pro-poor agricultural modernization refers to the advancement of agriculture technologies and institutions that improve the poor and small-scale farmers' welfare relative to rich large-scale farmers. This means that agricultural modernization includes mechanization strategy as part of technological change and the modernization of agriculture behavior, structure and institutions. The choice of the technology, which depends on the factor price and public policies, must be centered on the technological need of small-scale farmers.

The process of agriculture modernization includes mechanization and chemicalization. Mechanization comes with higher capital intensity whereas chemicalization implies that farmers adopt practices that increase the efficiency in the use of fertilizer and chemicals required to produce a certain level of outputs. This scheme includes also organic farming that maintains soil fertility to avoid the overuse of chemicals. Given the actual price and subsidies level, this technological path enables farmers to make effective and efficient use of the limited amount of fertilizer. These practices include crop rotation or integrated livestock crop rotation, intercropping, cover cropping or green manure, and composting waste materials. Finally, it is worth noting that achieving agricultural mechanization requires institutional changes that increase trust and encourage the private sector to adapt progressively proven technologies to local practices and production modes (Thirtle et al. [1998]). In the case of DRC, public interventions are required to improve access to markets for inputs, outputs and finance, as transaction costs are very high.

METHODS

To evaluate the pro-poorness of agricultural modernization-led growth, I adopt a sequential approach that combines a CGE model to a microsimulation model augmented to incorporate a pro-poor growth framework. The empirical strategy proceeds in four steps, as Fig. 7 depicts. I first use the CGE model to generate the effects of agricultural modernization strategies on employment, wages and rents, and the price of goods and services. Then I transmit these changes into a microsimulation model, which takes into account household heterogeneity in terms of factor endowments and consumption patterns, to generate welfare gains or losses at the household level. Using these changes in welfare, I apply the pro-poor growth framework to assess which of the agricultural modernization strategies is pro-poor, and the extent to which growth and redistribution contribute to welfare changes. Finally, I select a strategy that produced pro-poor welfare gains in the previous stage, and use a least square regression to explain its characteristics.

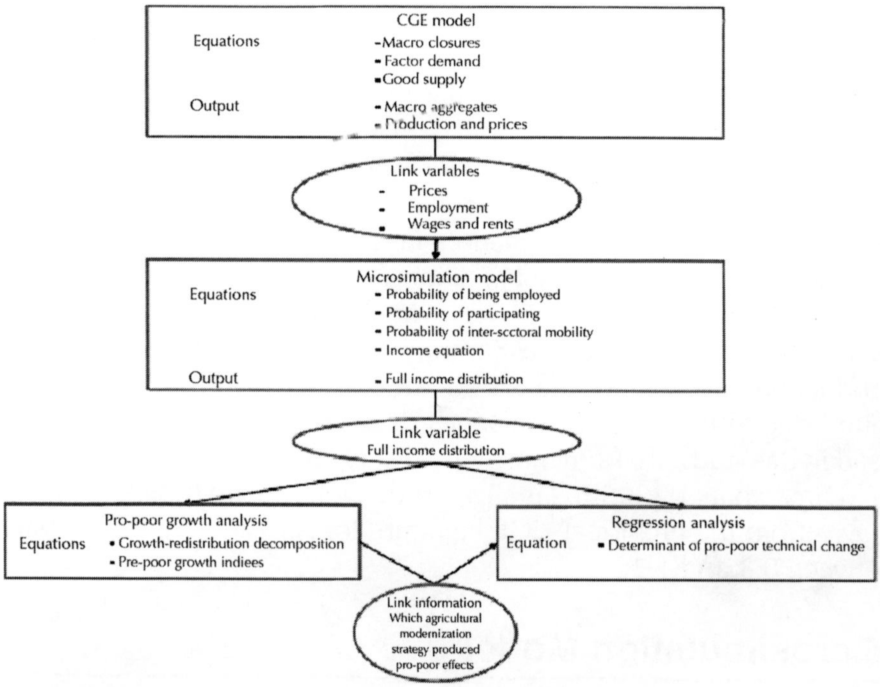

Figure 7: CGE-microsimulation augmented with pro-poor growth framework.

Congolese Computable General Equilibrium Model

The general specification of the Congolese CGE model follows the basic structure of the single-country model as described by Dervis et al. ([1982]). However, I closely follow Arndt et al. ([2000]) and Lofgren et al. ([2013]) for the specification of many structural and empirical features of the Congolese economy, namely an explicit modeling of trade and transportation costs for marketed commodities and relatively detailed description of home production. A full description of the CGE model can be found in Otchia ([2014]).

The Congolese CGE model is mainly calibrated to the social accounting matrix of the DRC for the year 2005 (Otchia [2013a]). However, additional data were required to fully run the CGE model, including household demand elasticities, trade elasticities, and production elasticities. Household demand elasticities include income elasticity and the Frisch parameter and were estimated based on the*Enquête* 1-2-3 survey data. Trade elasticities include elasticities for the Armington and transformation functions. Armington elasticities represent the elasticity of substitution in demand between imported commodities and domestic goods; whereas transformation elasticities include substitution elasticities among primary inputs in the value-added production function. For the case of the Congo, no trade elasticity was found due to the lack of time series data. Therefore, trade elasticities used in this study are from the Global Trade and Analysis Project based on Dimaranan ([2006]). Finally, production elasticities, which are drawn from the empirical CGE literature for African economies, vary between 0.3 and 1.2.

Microsimulation Model

The microsimulation model includes 12,098 households from the *Enquête* 1-2-3. My framework posits that agricultural technological change affects household income through channels such as changes in price of goods and services, changes in employment, and changes in the return to factors of production. The microsimulation model has two building blocks: a labor participation model and an accounting equation. Based on Magnac ([1991]) and Cogneau and Robilliard

([2001]), I specify a labor participation model to estimate changes in the labor conditions. Specifically, I use the labor participation equations to estimate the probability to participate in the labor market. Later, I use these probabilities to allocate labor in the microsimulation model based on changes in employment levels from the CGE model. The second component of the microsimulation transmits changes in commodity and factor prices following Otchia ([2014]).

The labor participation model has four components: (1) a probit model of the decision to participate in the labor market, (2) a multinomial probit model of the allocation of workers across sectors, (3) a bivariate probit model of the sectoral labor mobility, and (4) a rule for labor allocation and wage determination. The model assumes that workers can move from unemployment to employment status (or the opposite) and can move from across sectors. In the first stage, I estimate the choice of individuals to participate or not in the labor market. I run a probit model of employment to predict these probabilities, based on individual and households characteristics. The equation of the model is

$$\lambda_i = prob(I_i = 1 | z_i) = f(z_i \alpha + u_i) \quad (1)$$

where z_i represents individual and household characteristics of the household head such as age, gender, education, household composition; I_i is a binary variable which takes 1 if the household head is employed and 0 otherwise; and u_i is an error term. Similarly, I use a multinomial probit model to estimate the probability to be employed in each of the economic sectors, relative to the probability to be unemployed.

The second stage of the microsimulation model uses a bivariate probit model to estimate the decision of current workers to move from one sector to another. This model estimates the probability of workers to be employed in the new sector given their current employment status.

In the third step, I transmit employment levels taken from the CGE model into the microsimulation and determine which households are affected based on the job queuing approach (Bibi et al. [2010]). I rank the unemployed households by the decreasing order of their probability of being employed. Then I use the changes in employment from the CGE model to simulate the number of households who will be employed, starting from households with higher probability. The number of workers is calculated by multiplying the variation from the CGE model to sectoral employment and their labor income is the sectoral and skill level average. For sectors where employment shrinks, I rank employed households by the decreasing order to their probability of being unemployed. Given the changes from CGE model, I assign them the status of unemployed and their wage is set to zero.

When the demand for labor in one sector is higher than the supply from unemployment, I allow employed households to move to the sector with lack of supply based on their probability to move to other sectors. There is no cost of entry to the new sector as the CGE model assumed that labor moves to equalize wage.

In the final stage, I feed the changes in factor income (dw_l/w_l), marketed commodities prices (dp_g/p_g), and producer prices (dp_{hg}/p_{hg}) obtained from the CGE model into the accounting equation to determine welfare gains or losses of each of the 12,098 households. The first-order welfare change function dW_h/y_h is given as

$$dW_h / y_h = \sum_l \varnothing_h^l (dw_l / w_l) - \sum_g \theta_h^g (dp_g / p_g) - \sum_{hg} \theta_h^{hg} (dp_{hg} / ph_g) \qquad (2)$$

where $\varnothing_h^l$ is the share of factor *l* in factor income of household *h*, θ_h^g is the share of marketed good *g* in the total consumption expenditure of household *h*, θ_h^{hg} is the share of home-produced good *g* in the total consumption expenditure of household *h*, and y_h is the household income.

Growth-Redistribution Decomposition and Pro-Poor Growth Analysis

In this study, I apply the pro-poor growth framework on the welfare changes from the microsimulation to assess the pro-poorness of agricultural modernization strategies. I follow Kakwani and Pernia ([2000]) to decompose the total changes in welfare into two components: the pure growth effect and the pure inequality effect. Let L_B denote the distribution of income before agricultural modernization and L_A, the distribution of income after agricultural modernization. Then I write the growth rate in the mean income γ as

$$\gamma = \frac{\mu_A - \mu_B}{\mu_B} \tag{3}$$

where μ_B and μ_A are the mean of income before and after agricultural modernization. Thus, I define the pure growth effects *G* as the proportional change in welfare when the mean income changes but the distribution remains unchanged. The expression for the growth effects *G* is

$$G=P(\mu_A,L_A)-P(\mu_B,L_A) \tag{4}$$

or alternatively

$$G=P(\mu_A,L_B)-P(\mu_B,L_B) \tag{5}$$

Equivalently, the income effect depicts the change in welfare when inequality changes, but the mean income remains constant. This can be expressed as

$$R=P(\mu_B,L_A)-P(\mu_B,L_B) \tag{6}$$

or as

$$R=P(\mu_A,L_A)-P(\mu_B,L_A) \tag{7}$$

Following Kakwani ([2000]), I decompose the changes in welfare P^A-P^B as

$$P^A-P^B=0.5[P(\mu_A,L_A)-P(\mu_B,L_A)+P(\mu_A,L_B)-P(\mu_B,L_B)]+0.5[(\mu_B,L_A)-P(\mu_B,L_B)+P(\mu_A,L_A)-P(\mu_B,L_A)] \tag{8}$$

and then as

$$P^A - P^B = G + R \tag{9}$$

where G is the average growth effect, and R is the average income effect.

Based on this expression, Kakwani and Pernia ([2000]) introduced a pro-poor growth index to assess to what extent growth enables the poor to actively participate in and significantly benefit from it. This index is called the Kakwani and Pernia ([2000]) pro-poor index, Φ, and it is expressed as the ratio of the changes in poverty to the change that would have been observed if inequality did not change. Algebraically, this index is given by

$$\Phi = \frac{P^A - P^B}{G} \tag{10}$$

Depending on the values of γ and Φ, the Kakwani and Pernia ([2000]) index can be used to assess two types of pro-poor growth, namely, absolute pro-poor growth and relative pro-poor growth. Table 6 summarizes the decision matrix for the Kakwani and Pernia index.

Table 6: Decision matrix of Kakwani and Pernia index

		Decision
Positive	Positive	Absolute pro-poor growth
Positive	Negative	Absolute non pro-poor growth
Positive	Larger than 1	Relatively pro-poor growth
Positive	Lower than 1	Relatively non pro-poor growth

Source: Author

Otchia

Otchia *Journal of Economic Structures* 2014 3:8, doi: 10.1186/s40008-014-0008-x

Based on Kakwani and Pernia ([2000]) pro-poor index, Kakwani et al. ([2003]) further developed a pro-poor index, the Poverty Equivalent Growth Rate (PEGR) index, that adjusts for the change in the growth rate. The PEGR is thus defined as the growth rate that will result in the same observed level of poverty change had the actual growth process not been accompanied by any change in inequality. Algebraically, the PEGR is given by the product of Kakwani and Pernia ([2000]) pro-poor index (Φ) and the growth in average income (γ). It can, therefore, be written as

$$\gamma^{*} = \Phi\gamma. \tag{11}$$

According to this expression, growth is absolutely pro-poor if the PERG index is positive. Similarly, growth is relatively pro-poor if the PERG index exceeds the growth rate in average income. In the same way, Ravallion and Chen ([2003]) propose a measure of pro-poor growth using the mean growth rate of the poor as a measurement variable for the rate of pro-poor growth. They define the mean growth rate of the poor as the average growth in income of households below the poverty line. Thus, the Ravallion and Chen ([2003]) pro-poor index can be calculated as the mean of the growth rate of each percentile of the income distribution up to the headcount index, divided by the headcount index. This measure is equivalent to the actual growth rate multiplied by the ratio of the actual change in the Watts index to the change in the same index that would have occurred had growth been distribution neutral.

Determinants of Pro-Poor Growth

The pro-poor growth framework described in the previous section is highly descriptive and has limited policy implications. The reason is that this framework helps only to identify the sources of poverty changes, and policies that can produce pro-poor effects. In order to design a comprehensive policy package for agricultural transformation, I go a step further by ascertaining the relative contribution of specific factors that can potentially increase the pro-poorness of agricultural modernization. I proceed as follows. I select one of the agricultural modernization strategies that produced pro-poor effects, and run regressions on its welfare changes. The regression model includes variables that capture the heterogeneity of agriculture production, such as household socioeconomic characteristics, farm structure, and farming system. This is because the sole purpose of these regressions is to describe the profile of pro-poor technological changes at the household level. I estimate separate regressions for households in rural and urban areas, as the determinants of pro-poor welfare gains, as well as the policy implications, might be different in these two areas. The empirical specification of the pro-poor technological change, dW_h/y_h, is

$$dW_h/yh=\beta_0+\beta_1X_{1h}+\beta_2X_{2h}+\beta_3X_{3h}+\varepsilon_i, \quad (12)$$

where X_{1h} refers to household socioeconomic characteristics, X_{2h} represents the farm structure, and X_{3h} is a vector of other control variables. The vector of household socioeconomic characteristics includes household composition, the share of household members participating in off-farm activities, and some characteristics of the household head such as age and education. Farm structure is a vector of variables related to farm-labor relationship, gender, and access to land, farm tools, and access to credit. In the model, I allow for interaction effects between farm tools and access to credit (farm tools#credit). This simultaneous effect of credit and farm tools intuitively indicates the role of credit for purchasing farm machineries and implements. Finally, I use three variables to control for household heterogeneity at regional level. These include farming system, household index, and regional unemployment rate. Farming system is a binary variable which classifies regions according to the production of cassava, the leading crop in DRC. It takes a value of 1 if the region contributes less than 10 percent of total production of cassava, and 0 otherwise. Household index is a composite variable estimated by applying a factor analysis. It includes household assets and building materials, access to cleaned water, proximity to school and hospital, among others.

RESULTS AND DISCUSSION

This section experiments and analyzes agricultural modernization strategies using the methodology presented in the previous section. While the impacts of these policies on pro-poor growth will be the focus of this analysis, it is important that I discuss first the macroeconomic and sectoral effects of the experiments. Section 5.1 describes the policies considered in this study, Sects. 5.2 and 5.3 present the results of the CGE model on the macroeconomic and sectoral variables, and Sects. 5.4 and 5.5 present the pro-poor effects.

Policy Experiments

I design experiments to examine policy options for the DRC to modernize its agriculture and increase the income and welfare of households. In addition to their impact on income, I analyze if those policy experiments are pro-poor. As I noted in the theoretical consideration, there are two schemes to modernize agriculture in DRC. These are agricultural mechanization and chemicalization. On the order hand, I complement this analysis with two scenarios of institutional changes that lead to an increase in the supply of inputs. This is because institutional change is a key to the success of agricultural modernization in DRC as transaction costs within the input supply system are very high and there are no subsidies for inputs. Table 7 summarizes the policies simulated under the five scenarios.

Table 7: Design of policy experiments

Scenarios	Description	Design
Scenario 1	Capital-using technical change	Increase of capital intensity (3%) in agriculture
Scenario 2	Labor-using technical change	Increase of labor intensity (10%) in agriculture
Scenario 3	Improving the agricultural resource base (teach farmers how to better use fertilizer)	Reduction (10%) of chemicals per unit of aggregate intermediate input
		Reduction (20%) of factor requirements per unit of output
Scenario 4	Reducing fertilizer sourcing costs	Reduction (50%) of import trade margins for chemicals
		Reduction (50%) of domestic trade margins for chemicals

Scenario 5	Reducing fertilizer distribution costs	Reduction (50%) in import transportation costs for chemicals
		Reduction (50%) in domestic transportation costs for chemicals

Source: Author

Otchia

Otchia *Journal of Economic Structures* 2014 3:8, doi: 10.1186/s40008-014-0008-x

I conduct two sets of experiments to evaluate the pro-poorness of different agricultural modernization strategies. The first set of experiments focuses on technological change and includes three scenarios, of which two put emphasis on the mechanization policies that change the factor intensities in the production of agriculture goods, and one on the chemicalization policies that improve the efficiency in the use of chemicals. In Scenario 1, I assume that DRC adopts a capital-using technology in agriculture, increasing the capital intensity by 3 percent. Scenario 2 models the effects of a labor-using technology that increases the labor intensity by 10 percent. In Scenario 3, I implement a strategy for agriculture chemicalization. This scenario implies that DRC farmers apply organic chemicals and best practices to increase the effectiveness and efficiency of chemicals. This simulation also means that farmers invest in soil and water management to increase the effectiveness in the use of chemicals. Consequently, they can avoid the overuse of very expensive chemicals and increase their time or frequency on the farm. Technically, I implement this scenario as a joint simulation of a 10 percent reduction in the use of chemicals per unit of aggregate intermediate input and 20 percent increase in the factor requirement per agriculture output.

The second set of experiments considers a change in input markets related institutions. New institutions improve farmers' ability to access fertilizers and chemicals by reducing transaction costs in the procurement of inputs. The first experiment (Scenario 4) consists of reducing the fertilizers and pesticides sourcing costs by lowering imports and domestic trade margins by 50 percent. This can be achieved by cutting out the middlemen or breaking the monopoly in the distribution of agriculture inputs. Lastly, Scenario 5 models the effects of reducing fertilizers and pesticides transportation costs through the improvement of road and rail infrastructure and conditioning of products. In the model, I simulate the reduced fertilizer distribution costs scenario by assuming a 50 percent reduction in imports and domestic transportation costs of fertilizer.

In these scenarios, I assume that the direct tax rates are fixed and government savings adjust endogenously to ensure that government accounts balance. For the saving-investment account, the balanced closure specifies aggregate investment and government consumption as fixed shares of total domestic absorption. To keep investment fixed as a proportion of total absorption, the marginal savings propensities of households and enterprises move freely and maintain balance between investment and savings. Under this closure, agricultural modernization affects investment, consumption and savings evenly to absorb the full effect of the shock. Thus, the magnitude of the change of household consumption depends on the magnitude of the shock on savings, which in turn depends on the changes in investment and government consumption. The foreign exchange closure assumes a flexible exchange rate maintains a fixed level of foreign savings, which is consistent with the floating exchange rate system adopted in the DRC. Labor is unemployed and mobile across sectors, given high unemployment rate among all labor categories. This assumption is justified, as there is an excess supply of labor. In this case, real wage is fixed and the level of employment adjusts to restore the equilibrium in the labor market. Capital, in turn, is sector-specific but fully employed. Finally, the producer price index of non-traded domestic output is the numéraire.

Macroeconomic and Sector Effects

Table 8 reports the macro results of policy experiments described in the previous section. As expected, the impact of agricultural modernization is an increase in the size of the Congolese economy. Under capital-using technological change (Scenario 1), real GDP increases by 1.83 percent and total absorption by 1.80 percent. Because of the saving-investment balanced closure, the increased total absorption forces total investment and government consumption to rise by 0.59 and 0.58 percent. The increase of total investment depends also on the increase in government savings (1.99), as public investment accounts for 41 percent of total investment. At the same time, enterprise and household saving rates decrease by 0.17 and 1.08 percent, respectively, to match the value of investment. Private consumption increases by 2.04 percent while household and enterprise savings increase by 0.42 and 2.01 percent. Despite the fall in household and enterprise saving rate, savings by these domestic non-governmental institutions increase because of the impact of capital-using technological change on income. As a result of increased absorption, imports and exports rise, leading to an appreciation of the exchange rate.

Table 8: Impacts on macroeconomic variables

	Scenario 1	Scenario 2	Scenario 3	Scenario 4	Scenario 5
	Capital-using technical change	Labor-using technical change	Improved agricultural resource base	Reduced chemicals sourcing costs	Reduced chemicals distribution costs
GDP	1.83	2.19	1.00	0.46	0.35
Absorption	1.80	2.15	0.98	0.45	0.34
Private consumption	2.04	2.44	0.98	0.44	0.32
Government consumption	0.58	0.69	1.07	0.32	0.49
Total investment	0.59	0.69	1.00	0.52	0.40
Savings					
Enterprises	2.01	1.23	2.25	0.55	0.59
Households	0.42	2.32	3.71	0.37	0.97
Government	1.99	1.97	−1.07	−0.30	−0.29
Foreign sector	1.48	1.75	−0.18	−0.84	−0.37
Total	1.56	1.84	1.22	0.12	0.31
Marginal propensity to save					
Enterprises	−0.17	0.07	0.36	0.04	0.10
Households	−1.08	0.45	2.36	0.24	0.64
Exports	0.50	0.59	−0.34	0.92	0.25
Imports	0.46	0.55	−0.32	0.85	0.23
Exchange rate	1.48	1.75	−0.18	−0.84	−0.37

Consumer price index	0.28	0.32	0.18	−0.52	−0.05

Source: DRC CGE model

Otchia

Otchia *Journal of Economic Structures* 2014 3:8, doi: 10.1186/s40008-014-0008-x

Labor-using technological change (Scenario 2) leads to a 2.19 percent increase in GDP. Under this scenario, private and government consumption increase by 2.44 and 0.36 percent, respectively. Contrary to capital-using technological change, labor-using technological change would cause an increase in household saving rate by 0.45 percent and firm saving rate by 0.07 percent. From a policy perspective, the increase in household saving rate implies more private sector involvement in economic activity. In the case of agricultural households, savings can be invested in non-farming activities as well as in fertilizers and seeds. Finally, the simulation reveals that the real exchange rate appreciates by 1.75 percent, while exports and imports increase by 0.59 and 0.55 percent.

Scenario 3 indicates the potential impact of an improved agricultural resource base. The first thing to observe is that GDP and total absorption increase by 1 and 0.98 percent. As expected, total investment, household and government consumption also increase, due to the balanced closure. However, government savings decrease by 1.07 percent to balance government account because of fixed tax rates. Hence, only household and enterprise savings fuel total investment. Household savings increase significantly, reaching 3.71 percent, while firm savings increase by 2.25 percent. The remarkable change in household savings is mainly due to the increase in household savings rate (+2.36 %) as opposed to 0.36 percent for enterprises. Under this scenario, exports and imports fall by 0.34 and 0.32 percent, respectively.

A reduced chemical sourcing costs and reduced fertilizer distribution costs (Scenarios 4 and 5) yield quite similar macroeconomic results. However, the reduced chemicals sourcing costs scenario gives better impacts, mainly in terms of GDP, private consumption and total investment. The simulation exercise reveals that GDP increases by 0.46 and 0.35 percent in both scenarios. Further, reduced chemicals sourcing costs (Scenario 4) raises private consumption and total investment by 0.44 and 0.52 percent, while, under reduced chemicals distribution costs (Scenario 5), one can observe a 0.32 percent increase of private consumption and a 40 percent increase in total investment.

At a disaggregated level, capital-using and labor-using technological changes in agriculture (Scenarios 1 and 2) generate spillover and lead to the expansion of other sectors in the Congolese economy. Nevertheless, it is worth mentioning that agro-industry related sectors benefit less from agriculture technological change than manufacturing and service related sectors. For example, Table 9 indicates that under capital-using technological change the value added of processed food and textiles increases by 0.46 and 0.47 percent as opposed to 2.19, 1.40, and 0.73 percent for wood products, utilities, and education and health. The small expansion of agro-industries is due to weak forward linkages from agriculture and high transaction costs. This result reflects the dual characteristic of DRC economy where agriculture and service sectors have strong linkages and play a more active role in the economy.

Table 9: Changes in the value added

Sectors	Scenario 1	Scenario 2	Scenario 3	Scenario 4	Scenario 5
	Capital-using technical change	Labor-using technical change	Improved agricultural resource base	Reduced chemicals sourcing costs	Reduced chemicals distribution costs
Agriculture	6.92	8.31	0.27	0.03	0.02
Forestry	0.75	0.88	0.63	0.06	0.19
Mining	1.02	1.20	0.02	−0.48	−0.13
Processed foods	0.46	0.54	0.20	0.13	0.06
Textiles	0.47	0.56	−0.14	0.25	0.21
Wood	2.19	2.58	0.03	−1.53	−0.04
Chemicals	0.61	0.72	−1.36	0.83	0.48
Non-metals	0.68	0.80	0.64	0.07	0.30
Other manufacture	0.60	0.70	0.61	0.52	0.40
Utilities	1.40	1.66	0.88	0.01	0.42
Construction	0.57	0.67	0.93	0.46	0.37
Trade	0.33	0.39	−0.03	−1.00	0.32
Hotels and catering	0.73	0.86	0.01	0.24	0.17

Transportation	0.24	0.29	0.03	0.58	−0.37
Education and health	0.76	0.89	0.96	0.27	0.36
Other services	0.18	0.22	0.08	−0.05	0.04

Source: DRC CGE model

Otchia

Otchia *Journal of Economic Structures* 2014 3:8, doi: 10.1186/s40008-014-0008-x

Concerning improved agricultural resource base (Scenario 3), Table 9 shows that the contraction of the chemical sector is very pronounced compared to textile and trade. When I reduce chemical sourcing costs (Scenario 4), the value added of wood, trade, and mining decrease by 1.53, 1, and 48 percent. Similarly, reducing chemical distribution costs (Scenario 5) causes a contraction of the transportation (−0.37 %) and mining sectors (−0.13 %). Note that these two scenarios produce a very small increase in agricultural value added. Agricultural value added increases by around 0.03 percent under reduced chemicals sourcing costs and around 0.02 percent under reduced chemicals distribution costs.

Impacts on Factor Income and Consumption

Table 10 indicates the impact of agricultural modernization on factor income. Results show that capital-using technological change (Scenario 1) is skill-biased as it favors workers with higher skill. Since I allow for unemployment and sectoral labor mobility, changes in factor income are a response to the change in employment. This result indicates that mechanization reduces the demand for rural low-skilled and semi-skilled workers and increases the demand for rural high-skilled workers. In rural areas, capital and high skill appear to be complementary as capital-intensive technology increases the demand for high-skilled workers. High-skilled workers are needed to operate tractors or to spray chemical as manual work is less needed. Thus, the remuneration for rural unskilled and semi-skilled labor decreases by 3.12 and 1.35 percent, while rural workers with high skill levels see their remuneration increase by 1.74 percent.

Table 10: Impact on factor income

	Scenario 1	Scenario 2	Scenario 3	Scenario 4	Scenario 5
	Capital-using technical change	Labor-using technical change	Improved agricultural resource base	Reduced chemicals sourcing costs	Reduced chemicals distribution costs
Urban labor					
Unskilled	2.26	2.87	0.92	–0.79	0.52
Semiskilled	2.25	2.70	0.45	–0.16	0.27
Skilled	2.25	2.66	0.24	0.07	0.31
Rural labor					
Unskilled	–3.12	8.48	1.38	–0.48	0.29
Semiskilled	–1.35	6.33	1.14	–0.63	0.27
Skilled	1.74	4.35	1.25	–0.55	0.59
Capital	1.85	1.09	1.47	0.25	0.34
Land	1.37	0.87	1.31	0.38	0.45

Source: DRC CGE model

Otchia

Otchia *Journal of Economic Structures* 2014 3:8, doi: 10.1186/s40008-014-0008-x

Labor-using technological change (Scenario 2) influences labor income by increasing the intensity of workers on the farm and, therefore, raising working hours and the frequency of work throughout the year. Scenario 2 in Table 10 indicates that labor-using technological change raises the returns for all factors involved in the production process. As expected, rural workers capture higher benefits. However, it is interesting to see that many of the gains go to workers with lower skills. This is because agriculture uses low-skilled workers intensively. In addition, this scenario has important implications for income distribution in both rural and urban areas, as gain differences in favor of workers with lower skill are non-negligible. For example, the remuneration of rural low-skilled workers increased by 8.48 percent whereas the return for semi-skilled and high-skilled laborers increased by 6.33 and 4.35 percent.

With regard to the improved agricultural resource base scenario, one can see that the results are qualitatively similar with labor-using technological change. Returns to labor are higher among rural workers than urban workers with similar skills. Under this scenario, urban low-skilled workers gain more than urban semi-skilled and high-skilled workers do. Table 10 shows that returns to urban low-skilled labor increased by 0.92 percent while the returns for semi-skilled and high-skilled workers increased by 0.45 and 0.24 percent, respectively.

Turning the attention to the institutional changes scenarios, simulation results show that reducing fertilizer sourcing costs (Scenario 4) decreases the factor income for all labor categories, except for urban high-skilled labor. Reducing trade margins for fertilizers and pesticides lowers the price for agriculture intermediate inputs and increases agriculture producer price. Rising producer price in agriculture, in principle, should lead to an increase in factor income of rural workers as they are intensively employed in agriculture. However, the simulation indicates the fall of factor income. Income for urban low-skilled and semi-skilled labor falls by 0.79 and 0.16 percent, respectively. In rural areas, the remuneration for unskilled and semi-skilled labor drops by 0.48 and 0.65 percent, respectively, while high-skilled workers see their remuneration fall by 0.55 percent. There are two reasons for this. First, the impact of reduced fertilizer sourcing costs on agriculture value added is marginal, as agriculture uses outdated technology. From a policy standpoint, this indicates that technological changes play an important role in boosting agriculture output and productivity. Second and most importantly, lowering trade margins leads to a reduction of the producer price in the trade sector. This is due to the adjustment costs that occur when the Congolese marketing system transforms to a modern sector. Reduced producer price lowers the value-added price and thus wages of most of the labor categories since this sector is the second employer after agriculture. To counterbalance those negative effects on labor income, policies that break the monopoly of intermediaries in agricultural trade should also aim to increase operational efficiency by focusing on reducing the costs of inputs. This can be done when DRC public policymakers aim to improve ICT services to farmers by removing administrative barriers that prevent the development of mobile banking, mobile remittance and the exchange of agricultural input price information.

If improving institutions lowers the fertilizer distribution costs, as in Scenario 5, then the income of all production factors will increase. Under this scenario, the changes in income of rural workers are higher than those of urban areas. The highest increase is attributed to unskilled rural workers whose income increases by 1.38 percent whereas urban high-skilled labor's income increases only by 0.24 percent. Meanwhile, returns to capital and land are also high and reach 1.47 and 1.31 percent, respectively.

Moving on to the consumption effects of agriculture products, Table 11 indicates that mechanization of agriculture (Scenarios 1 and 2), leads to a decrease of the price of agricultural products and an increase in the competitiveness of domestic products. Improving the agricultural resource base (Scenario 3) raises the market price of agricultural products, but this leads to an insignificant fall in sales. On the other hand, lowering chemicals sourcing costs (Scenario 4) reduces the market price of domestic production marginally, whereas reduced chemicals distribution costs (Scenario 5) increases the market price. Looking at consumption changes in Table 11, one can see that agricultural mechanization (Scenarios 1 and 2) increases household consumption of agricultural products from the market and their own production. Changes from home-produced consumption are significantly higher than the changes in consumption from the market. This is due to the high rate of subsistence agriculture and significant price differences due to transaction costs. It is interesting to see that in the case of an improved agricultural resource base (Scenario 3), home-produced consumption increases to a lesser extent than marketed consumption. Finally, lowered chemical sourcing costs (Scenario 4) reduces the consumption from own production and increases consumption from markets. In this scenario, home-produced consumption decreases by 0.12 percent while marketed consumption increases by 0.37 percent. For the reduced chemicals distribution costs scenario (Scenario 5), one can see that agriculture consumption from market and home production increases marginally. However, it is important to note that market consumption increases more than home-produced consumption.

Table 11: Impacts on consumption of agricultural products

	Scenario 1	Scenario 2	Scenario 3	Scenario 4	Scenario 5
	Capital-using technical change	Labor-using technical change	Improved agricultural resource base	Reduced chemicals sourcing costs	Reduced chemicals distribution costs
Price for domestically produced and sold goods	−6.32	−7.46	0.15	−0.09	0.25
Price of composite goods	−5.77	−6.82	0.12	−0.20	0.23
Quantity of domestic sales	5.21	6.23	−0.05	0.12	0.04
Quantity of aggregate marketed output	5.49	6.58	−0.06	0.10	0.03
Quantity of marketed consumption	7.99	9.55	1.13	0.37	0.06
Quantity of home consumption	10.10	12.11	0.99	−0.12	0.01

Source: DRC CGE model

Otchia

Otchia *Journal of Economic Structures* 2014 3:8, doi: 10.1186/s40008-014-0008-x

Pro-Poor Growth and Growth-Redistribution Decomposition

In order to understand how inclusive the different schemes of technological and institutional changes are, in this section I apply the framework of pro-poor growth analysis on the welfare gains from the CGE results. Table 12 gives the estimates of the growth in average income and six pro-poor indices. The pro-poor indices include the Ravallion and Chen ([2003]) index, the Ravallion and Chen ([2003])

index minus γ, the Kakwani and Pernia ([2000]) index, the Kakwani and Pernia ([2000]) index minus 1, the poverty-equivalent growth rate (PEGR) index, and the poverty-equivalent growth rate (PEGR) index minus γ. The Ravallion and Chen index, the Kakwani and Pernia index, and PEGR index constitute the absolute pro-poorness indices. They indicate whether the income of the poor has grown sufficiently after agricultural modernization for absolute poverty indices to fall. Therefore, a positive value of these indices indicates that growth led by agricultural modernization is absolutely pro-poor. The other three indices, the Ravallion and Chen index minus γ, the Kakwani and Pernia index minus 1, and PEGR index minus γ, depicts the relative pro-poorness of agricultural modernization. They demonstrate whether the income of the poor has grown sufficiently after agricultural modernization to follow the overall increase in average income (γ). In this case, agricultural modernization is relatively pro-poor if these indices take positive values.

Table 12: Pro-poor analysis

Indices	Scenario 1	Scenario 2	Scenario 3	Scenario 4	Scenario 5
	Capital-using technical change	Labor-using technical change	Improved agricultural resource base	Reduced chemicals sourcing costs	Reduced chemicals distribution costs
Growth rate ( )	3.06	4.89	0.42	0.28	0.15
Ravallion and Chen ([2003])	−0.24	3.31	0.28	0.03	0.12
Ravallion and Chen ([2003]) –	−3.30	−1.58	−0.14	−0.25	−0.04
Kakwani and Pernia ([2000])	−0.16	1.89	1.25	0.07	0.90
Kakwani and Pernia ([2000]) – 1	−1.16	0.89	0.25	−0.93	−0.10

PEGR	−0.49	8.29	0.53	0.02	0.14
PEGR −	−3.55	3.40	0.11	−0.26	−0.02

Source: Author's calculation

Otchia

Otchia *Journal of Economic Structures* 2014 3:8, doi: 10.1186/s40008-014-0008-x

Table 12 shows that all the three scenarios of agriculture modernization (Scenarios 1, 2, and 3) lead to an increase in average income. This result is consistent with the macro effects discussed in the previous section. Table 12 further indicates that capital-using technological change (Scenario 1) is neither absolutely pro-poor nor relative pro-poor, as all of the six indices are negative. This implies that agriculture growth led by capital-using technological change decreases significantly the poor's relative shares in total consumption. In contrast, labor-using technological change (Scenario 2) is absolutely pro-poor. This is indicated by the positive sign of the Ravallion and Chen index, the Kakwani and Pernia index, and PEGR index. In terms of relative pro-poor effects, the Kakwani and Pernia index minus 1 and the PEGR index minus γ show that the growth rate of the poor's income is enough to follow the growth rate in average income. This gives evidence that labor-using technical change-led growth is relatively pro-poor.

An improved agricultural resource base (Scenario 3), on the other hand, is absolutely pro-poor as all the three indices are greater than zero. Looking at this in a relative perspective, one can find that this policy is also pro-poor as the two relative pro-poor indices are positive. Turning our attention to the reduced chemical sourcing costs (Scenario 4), the absolute pro-poor indices are positive but not significant. However, the results on the relative pro-poor indices indicate that reduced chemical sourcing costs are not relatively pro-poor. Finally, the last column of Table 12 presents the pro-poorness of reduced chemical distribution costs (Scenario 5). The findings indicate that this scenario is absolutely pro-poor, but I have little evidence to conclude about the relative pro-poorness.

Next, I am interested in understanding the source of changes in poverty due to agricultural modernization. For this purpose, I decompose changes in poverty headcount ratio in terms of the effect of growth and changes in redistribution. The first column of Table 13 presents the growth-redistribution decomposition of the impact of capital-using technological change (Scenario 1). Under this policy scenario, poverty headcount ratio increases by 2.07 percent. However, it is interesting to see that without any changes in inequality, capital-using technological change would reduce poverty by 1.05 percent. The increase in inequality (+3.15 %) cancels out the beneficial effect of capital-using technological change-led growth on poverty reduction. Thus, capital-using technological change leads to immiserizing growth. This finding corroborates the ideas that poverty reduction due to growth led by large-scale investment in agriculture depends on the initial level of inequality in income or distribution of assets. This is in line with previous research (Bourguignon and Morrisson[1998]; de Janvry and Sadoulet [1996]; Timmer [1997]; Ravallion [1997]; World Bank [2000]), suggesting that the distribution of assets matters as it affects how well the poor connect to the growth process.

Table 13: Growth-redistribution decomposition

	Scenario 1	Scenario 2	Scenario 3	Scenario 4	Scenario 5
	Capital-using technical change	Labor-using technical change	Improved agricultural resource base	Reduced chemicals sourcing costs	Reduced chemicals distribution costs
Changes in poverty index	2.07	−3.47	−2.71	−0.13	−1.13
Growth	−1.05	−2.91	−1.76	−2.55	−1.15
Redistribution	3.12	−0.56	−0.94	2.43	0.02

Source: Author's calculation

Otchia

Otchia *Journal of Economic Structures* 2014 3:8, doi: 10.1186/s40008-014-0008-x

In contrast, labor-using technological change (Scenario 2) causes a reduction in the poverty headcount by 3.47 percent. 2.91 percent points of the 3.74 percentage point fall in poverty headcount are due to the growth effect. This means that if inequality did not change, poverty would be reduced by 2.91 percent. Thus, redistribution was responsible for 0.56 percentage points of poverty reduction. These findings are consistent with earlier evidence that unskilled labor-intensive agricultural activities have higher poverty-reducing capacity compared to high-skilled, capital-intensive activities (de Janvry and

Sadoulet [2010]; Loayza and Raddatz [2010]). Similarly, an improved agricultural resource base (Scenario 3) reduces poverty by 2.71 percent. The growth effect contributes to 1.76 percentage points in poverty changes while the income distribution effect contributes to 0.94 percentage point.

Concerning the growth-redistribution decomposition of reduced chemicals sourcing and distribution costs (Scenarios 4 and 5), one can observe in Table 13 that these scenarios have the same qualitative effects. They both lead to poverty reduction as would be expected. The growth effects of these policy scenarios contribute to reducing poverty whereas the redistribution effects contribute to increasing poverty. Nonetheless, the contribution of the redistribution effects is very low under reduced chemicals distribution costs compared to reduced chemical sourcing costs.

Explaining Pro-Poor Technological Change

I extend the pro-poor growth analysis by looking at the determinants of pro-poor technological change at household level. Recall from the previous section, I found that labor-using technological change (Scenario 2) is absolutely and relatively pro-poor. Therefore, I run regressions on its predicted welfare gains to ascertain the relative contribution of relevant observed characteristics that can potentially increase the pro-poorness of technological change. Table 14 gives summary statistics on the predictors used in the regressions, broken down in rural and urban areas. In order to reduce any potential endogeneity problems of some of the predictors, I choose to include in the regressions only explanatory variables of potential relevance to agricultural and rural development policies in DRC. The first set of variables consist of household characteristics such household composition and the head's education and age. I also use the share of household members participating in off-farm activities as a proxy to measure the importance of off-farm activities and thus household income diversification. The second group of predictors is made up of farm structure variables such as farm-labor relationship, access to credit, farm tools possession, and rights on land. Finally, I use farming system, household index, and regional unemployment rate to control household heterogeneity at regional level. I define these predictors such that a positive sign implies better pro-poor welfare gains.

Table 14: Summary statistics on predictors in the regression analysis

Variable	Rural		Urban		National	
	Mean	Standard deviation	Mean	Standard deviation	Mean	Standard deviation
Household composition						
Household size (log)	1.576	0.62	1.493	0.60	1.531	0.61
Share of kids in the household	0.181	0.18	0.195	0.18	0.189	0.18
Share of young in the household	0.454	0.25	0.460	0.25	0.457	0.25
Share of adults in the household	0.505	0.25	0.493	0.25	0.499	0.25
Participation in off-farm activities (share)	0.578	0.28	0.888	0.21	0.720	0.29
Age of household head (log)	3.741	0.31	3.699	0.33	3.718	0.32
Squared age (log)	14.095	2.33	13.794	2.44	13.932	2.40
Years of education (log)	2.099	0.53	1.758	0.59	1.923	0.59
Squared years of education (log)	4.687	1.90	3.438	1.83	4.043	1.96
Farm structure						
Farm-labor relationship						
Household head or spouse	0.250	Binary	0.632	Binary	0.457	Binary
Other household members	0.025	Binary	0.027	Binary	0.026	Binary
Wage workers	0.005	Binary	0.008	Binary	0.006	Binary
Sharecropper	0.000	Binary	0.002	Binary	0.001	Binary
Other	0.003	Binary	0.003	Binary	0.003	Binary
Male head with spouse without rights on land	0.803	Binary	0.834	Binary	0.820	Binary

Female head holding rights on land	0.003	Binary	0.001	Binary	0.002	Binary
Male head with spouse with rights on land	0.004	Binary	0.007	Binary	0.005	Binary
Farm tools	0.816	Binary	0.985	Binary	0.908	Binary
Credit	0.092	Binary	0.118	Binary	0.106	Binary
Regional characteristics						
Farming system	0.333	Binary	0.154	Binary	0.236	Binary
Household index	0.835	1.10	−0.382	0.47	0.174	1.02
Unemployment rate	5.654	4.68	3.539	2.30	4.506	3.74

Source: Author's calculation

Otchia

Otchia *Journal of Economic Structures* 2014 3:8, doi: 10.1186/s40008-014-0008-x

Table 15 shows the results on the determinants of pro-poor technological change derived from the regression model outlined in Eq. (12). They indicate that pro-poor welfare gains increase with household size in both rural and urban areas. However, the results suggest a strong negative and significant relationship between pro-poor welfare gains and household composition, especially concerning younger household members. For example, I find that rural and urban households with a larger share of kids tend to have lower welfare gains. This implies that the number of children also affects women's choice to work on the farm, as women are responsible for most of the on-farm tasks. Interestingly, I find that participation in off-farm activities is positively correlated with pro-poor welfare gains in rural areas and negatively in urban areas. This indicates that participation in off-farm activities has positive spillover effects on pro-poor agricultural technological change in rural areas. An important reason for this is that in rural areas, off-farm income is usually invested in modern inputs and insurance. This finding is in line with other studies that found that investment in non-farm activities can benefit the agricultural sector (Dorward et al. [2004]; de Janvry et al. [2005]).

Table 15: Explaining pro-poor technological change led growth

Variables	Rural	Urban
Household composition		
Household size (log)	0.262***	0.327***
Share of kids in the household	−0.216***	−0.308***
Share of young in the household	−0.463***	−0.021
Share of adults in the household	−0.381***	0.112
Participation in off-farm activities (share)	0.219***	−0.181***
Age of household head (log)	7.549***	7.029***
Squared age (log)	−1.030***	−0.916***
Years of education (log)	0.007	−0.237*
Squared years of education (log)	−0.010	0.098**
Farm structure		
Farm-labor relationship		
Household head or spouse (reference)		
Other household members	0.027	0.054
Wage workers	0.415***	−0.060
Sharecropper	0.213	0.813
Other	0.174	−0.472*
Male head with spouse without rights on land	0.100***	0.137***
Female head holding rights on land	0.004	0.351
Male head with spouse with rights on land	0.183*	0.454*
Farm tools	0.124	−0.543
Credit	0.166**	0.495***
Farm tools#credit (interaction term)	0.171**	0.571***
Regional characteristics		
Farming system	−2.606***	−0.660***
Household index	0.189***	0.133***
Unemployment rate	0.452***	0.127***

Observations	4767	2192
Adjusted ρ^{P}	0.898	0.897

$^{*}p<0.10$, $^{**}p<0.05$, $^{***}p<0.01$

Source: Author's estimations

Otchia

Otchia *Journal of Economic Structures* 2014 3:8, doi: 10.1186/s40008-014-0008-x

In addition, I found that pro-poor welfare gains are inverted U-shaped in age of household head in both rural and urban areas. This indicates that age has diminishing returns, meaning that it is beneficial for pro-poor growth until 39 in rural areas and 46 in urban areas, after which increases in age will decrease pro-poor welfare gains. The results for the years of education of the head are mixed. I find that education of the head has a very small and non-significant inverted U-shape effect in rural areas, but a significant U-shape effect in urban areas. To clearly highlight the substantive significance of education, I estimate and present in Fig. 8 the predictive margins for the years of education of the head in both rural and urban areas. Perhaps most striking from Fig. 8, compared to the sign in Table 15, is the steady decline in pro-poor welfare gains until approximately 5 years of education. From a policy perspective, this finding means that every advance in post-primary education leads to higher pro-poor welfare gains.

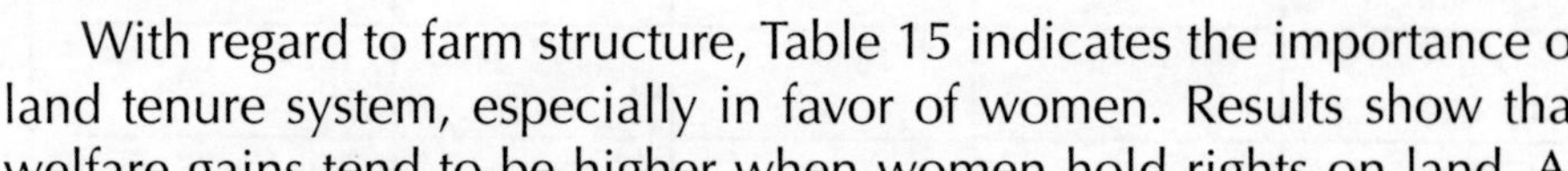

Figure 8: Predicative margins of education of the head.

With regard to farm structure, Table 15 indicates the importance of land tenure system, especially in favor of women. Results show that welfare gains tend to be higher when women hold rights on land. As can be seen, welfare gains increase by 0.183 percent when the head of household is a male and the spouse holds rights on the land. When the spouse does not hold rights on land, the increase in welfare gains is only 0.1 percent. It should be worth mentioning that this result is consistent in both rural and urban areas, with a higher magnitude for urban areas. Further, farm tools have opposite sign in rural and urban areas but I fail to find any significant effect. However, what is interesting to notice from the estimation results is that the interaction between farm tools and credit is positive and significant. It is also worth mentioning that the coefficient of the interaction between farm tools and credit is larger than the effect of credit alone. Intuitively, this indicates the importance of establishing a specific credit for purchasing farm tools.

Sensitivity Analysis

A common feature of CGE model results is that they depend on assumptions made. In this section, I conduct a sensitivity analysis of the CGE model to ensure the robustness of the results. In the sensitivity analysis, I show the comparison for the simulation of capital-using and labor-using technical change with respect to the change of the production factor elasticities. Table 16 presents some of the results obtained when production elasticities of substitution increase or decrease by 5 percent.

Table 16: Explaining pro-poor technological change led growth

	Capital-using technical change		Labor-using technical change	
	ρ^P decreases by 5 %	ρ^P increases by 5 %	decreases by 5 %	increases by 5 %
GDP	2.61	0.37	2.66	1.80
Absorption	2.57	0.37	2.61	1.77
Private consumption	2.92	0.41	2.97	1.97
Government consumption	0.77	0.17	0.78	0.77
Total investment	0.78	0.16	0.79	0.74
Exports	0.67	0.14	0.68	0.67
Imports	0.62	0.13	0.63	0.63

Otchia

Otchia *Journal of Economic Structures* 2014 3:8, doi: 10.1186/s40008-014-0008-x

I expected the magnitude of the results to differ because the nature of technical change and the assumption on the elasticities of substitution are the main drivers of model results. Nonetheless, the qualitative results and the signs are robust to the changes in elasticities. Nevertheless, it is worth mentioning that the capital-using simulation appears to be less robust to higher production elasticities of substitution.

CONCLUSIONS AND POLICY IMPLICATIONS

In analyzing what are the better or worse models for agriculture modernization, this paper found that the adoption of capital-using technology leads to immiserizing growth through the redistribution effects. Results indicate that growth effects under this technological change reduced poverty slightly, while the redistribution effects canceled out the positive growth effects. This is because capital-using technological change increased output in all the sectors, but lowered rural unskilled and semi-skilled income so that there was an overall increase in inequality. Rising inequality between rural workers with lower skill and the rest of workers emerged, a key factor explaining the anti-poor effects of capital-using technological change.

In contrast, labor-using technological change is found to be effective in producing pro-poor effects in both absolute and relative terms. The analysis also pinpoints the importance of labor-using technological change in improving urban-rural income disparities. Despite the large income change in favor of rural workers with lower skills, this result is partly due to the increase of home-produced consumption relative to the marketed consumption. The findings on poverty decomposition demonstrate that growth was responsible for more of the poverty changes than redistribution. This finding suggests that labor-using technological change can be independently sufficient in reducing poverty via the income growth effects. Under this scenario, I found that household and firm savings also increased. This can be an indication of future private investment in non-farm or fertilizer related activities. The improved agricultural resource base scenario produced similar pro-poor effects to those in labor-using technological change. Nonetheless, improved agricultural resource base has a limited potential to enhance growth outside agriculture.

In addition, I tested two scenarios of institutional changes that lead to an increase in supply of agriculture inputs. Firstly, the reduction of chemical sourcing costs implies lower income for most of the labor types and produces insignificant absolute pro-poor gains. From a policy perspective, this finding suggests that reducing trade margins should be implemented simultaneously with institutions and policies that increase farmers' market power and improve marketing efficiency. Secondly, reducing chemicals' distribution costs increases access to market for both producers and consumers, and it leads to an increase in income for all labor types. Consistently, this policy scenario produces absolute pro-poor effect but does not lead to income convergence.

I continued the analysis by investigating quantitatively the determinants of pro-poor growth, using welfare changes from labor-using technological change. Doing so, I found that participation in off-farm activities is statistically significant and strong determinant of pro-poor technological change in agriculture. Working as a wage worker improves pro-poor welfare gains only in rural areas. Other important findings show that women's land rights emerged as an important determinant of pro-poor welfare gains. Credit is positive and significant in rural and urban areas but I have not found significant effects for farm tools. Nonetheless, I found that there exist significant interactions between access to credit and farm tools in producing pro-poor welfare gains.

This research has intuitive findings for design and implementation of a pro-poor agriculture modernization strategy. The key policy recommendations arising from the paper are the following:

1. Public policymakers should promote the adoption of labor-using technologies to enable the use of cheap labor to intensify agriculture.
2. Public policymakers should increase farmers' capacity to evaluate, adapt, and disseminate proven technologies.
3. Public policymakers should increase investment in soil and water management methods, and in agriculture research and extension, to improve farmer's ability to use fertilizer efficiently.
4. Public policymakers should secure tenure among small-scale farmers and improve access to land, especially for women.
5. Public policymakers should help farmers organize themselves into cooperatives, break monopolies and cut rent seekers in seed

supply and increase marketing efficiency.

6. Public policymakers should reform input supply networks and increase investment in input storage and road infrastructure. These steps should be implemented together with policies and institutions to remove credit constraint and increase farmers' ability to acquire tools and inputs.

ACKNOWLEDGEMENTS

The author would like to thank the anonymous referees for their comments and suggestions.

REFERENCES

1. (2012) Land policy in Africa: Central Africa regional assessment. AUC-ECA-AfDB Consortium, Addis Ababa.
2. Alvarez-Cuadrado F, Poschke M (2011) Structural change out of agriculture: labor push versus labor pull. Am Econ J Macroecon 3(3):127-158
3. Annabi N, Cissé F, Cockburn J, Decaluwé B (2008) Trade liberalisation, growth and poverty in Senegal: a dynamic microsimulation CGE model analysis. Econ Prévis 186(5):117-131
4. Arndt C, Robinson S, Tarp F (2000) Marketing margins and agricultural technology in Mozambique. J Dev Stud 37(1):121-137
5. Bibi S, Cockburn J, Tiberti L, Fofana I (2010) Impacts of the global crisis and policy responses on child well-being: a macro–micro simulation framework. Innocenti working paper 2010-06, UNICEF Innocenti Research Centre, Florence and UNICEF Regional Office for West and Central Africa, Dakar
6. Boccanfuso D, Kaboré ST (2004) Croissance, inégalité et pauvreté dans les années quatre-vingt-dix au Burkina Faso et au Sénégal. Rev écon Dév 2(12):9-35
7. Boccanfuso D, Coulibaly M, Timilsina GR, Savard L (2011) Economic and distributional impacts of biofuels in Mali. Cahiers

de recherche 11-08, Departement d'Economique de la Faculte d'administration à l'Universite de Sherbrooke

8. Boccanfuso D, Coulibaly M, Timilsina GR, Savard L (2013a) Macroeconomic and distributional impacts of jatropha-based biodiesel in Mali. Policy research working paper 6500, The World Bank, Washington

9. Boccanfuso D, Savard L, Estache A (2013) The distributional impact of developed countries' climate change policies on Senegal: a macro–micro CGE application. Sustainability 5(6):2727-2750

10. Bourguignon F, Morrisson C (1998) Inequality and development: the role of dualism. J Dev Econ 5:233-257

11. Cameron AC, Trivedi PK (2010) Microeconometrics using Stata. Stata Press, College Station.

12. Chausse J-P, Kembola T, Ngonde R (2012) L'agriculture : pierre angulaire de l'économie de la RDC. In: Herderschee J, Samba DM, Tshibangu MT (eds) Résilience d'un Géant Africain : Accélérer la Croissance et Promouvoir l'Emploi en République Démocratique du Congo, Volume II : Etudes sectorielles, MÉDIASPAUL, Kinshasa. pp 1-97

13. Chen S, Ravallion M (2004) Welfare impacts of China's accession to the World Trade Organization. World Bank Econ Rev 18(1):29-57

14. Cleaver HM Jr (1972) The contradictions of the Green Revolution. Am Econ Rev 62(1/2):177-186

15. Cogneau D, Robilliard A-S (2001) Croissance, distribution et pauvreté : un modèle de micro simulation en équilibre général applique à Madagascar. Working paper DT/2001/19, DIAL (Développement, Institutions & Analyses de Long terme)

16. Collier P, Soludo CC, Pattillo C (2008) Economic policy options for a prosperous Nigeria. Palgrave Macmillan, New York.

17. Cuffaro N (1997) Population growth and agriculture in poor countries: a review of theoretical issues and empirical evidence. World Dev 25(7):1151-1163

18. Das RJ (1998) The Green Revolution, agrarian productivity and labor. Int J Urban Reg Res 22(1):122-135

19. Dawkins C, Srinivasan TN, Whalley J (2001) Calibration. In: James JH, Edward L (eds) Handbook of econometrics, Elsevier, Amsterdam. pp 3653-3703 (chap 58)

20. de Janvry A, Sadoulet E (1996) Growth, inequality and poverty in Latin America: a causal analysis, 1970–94. Working paper 784, Department of Agricultural and Resource Economics, University of California at Berkeley
21. de Janvry A, Sadoulet E (2010) Agricultural growth and poverty reduction: additional evidence. World Bank Res Obs 25(1):1-20
22. de Janvry A, Sadoulet E, Zhu N (2005) The role of non-farm incomes in reducing rural poverty and inequality in China. CUDARE working paper series 1001, Department of Agricultural and Resource Economics and Policy, University of California at Berkeley
23. Dervis K, de Melo J, Robinson S (1982) General equilibrium models for development policy. Cambridge University Press, New York.
24. DFID (2006) Technology and its contribution to pro-poor agricultural development. DFID working paper 4, University of Greenwich
25. Dimaranan BV (ed) (2006) Global trade, assistance, and production: the GTAP 6 Data Base. Center for Global Trade Analysis, Purdue University
26. Dorward A, Fan S, Kydd J, Lofgren H, Morrison J, Poulton C, Rao N, Smith L, Tchale H, Thorat S, Urey I, Wobst P (2004) Institutions and policies for pro-poor agricultural growth. Dev Policy Rev 22(6):611-622
27. Eicher CK (1995) Zimbabwe's maize-based Green Revolution: preconditions for replication. World Dev 23(5):805-818
28. FAO/WHO/UNU (2004) Human energy requirements. Report of a Joint FAO/WHO/UNU Expert Consultation, FAO food and nutrition technical report series 1
29. Freebairn D (1995) Did the Green Revolution concentrate incomes? A quantitative study of research reports. World Dev 23:265-279
30. Glaeser B (1987) The Green Revolution revisited. Allen & Unwin, London.
31. Goldman A, Smith J (1995) Agricultural transformations in India and Northern Nigeria: exploring the nature of Green Revolutions. World Dev 23(2):243-263

32. Griffin K (1979) The political economy of agrarian change: an essay on the Green Revolution. Macmillan Press, London.

33. Hayami Y, Ruttan VW (1970) Factor prices and technical change in agricultural development: the United States and Japan, 1880–1960. J Polit Econ 78(5):1115-1141

34. Hayami Y, Ruttan VW (1985) Agricultural development: an international perspective. Johns Hopkins University Press, Baltimore.

35. Hazell P (2009) The Asian Green Revolution. IFPRI discussion paper 00911, Washington

36. Hazell P, Ramasamy C (1991) The green revolution reconsidered: the impact of high-yielding rice varieties in South India. Johns Hopkins University Press, Baltimore.

37. Herderschee J, Kaiser K-A, Mukoko DS (2012) Resilience of an African giant: boosting growth and development in the Democratic Republic of Congo. The World Bank, Washington.

38. Huggins C (2010) Land, power and identity: roots of violent conflict in Eastern DRC. International Alert, London.

39. Irz X, Lin L, Thirtle C, Wiggins S (2001) Agricultural productivity growth and poverty alleviation. Dev Policy Rev 19(4):449-466

40. Jayne TS, Yamano T, Weber MT, Tschirley D, Benfica R, Chapoto A, Zulu B (2003) Smallholder income and land distribution in Africa: implications for poverty reduction strategies. Food Policy 28(3):253-275

41. Kakwani N (2000) On measuring growth and inequality components of poverty with application to Thailand. J Quant Econ 16(1):67-80

42. Kakwani N, Pernia EM (2000) What is pro-poor growth? Asian Dev Rev 18(1):1-16

43. Kakwani N, Khandker S, Son HH (2003) Poverty equivalent growth rate: with applications to Korea and Thailand. Technical report, Economic Commission for Africa

44. Lambert PJ, Aronson JR (1993) Inequality decomposition analysis and the Gini coefficient revisited. Econ J 103(420):1221-1227

45. Lambert PJ, Decoster A (2005) The Gini coefficient reveals more. Discussion paper series (DPS) 05.08, Center for Economic Studies

46. Loayza NV, Raddatz C (2010) The composition of growth matters for poverty alleviation. J Dev Econ 93(1):137-151
47. Lofgren H, Cicowiez M, Diaz-Bonilla C (2013) MAMS—a computable general equilibrium model for developing country strategy analysis. In: Dixon PB, Jorgenson D (eds) Handbook of computable general equilibrium modeling, Elsevier, Amsterdam.
48. Long C (2011) Land rights in the Democratic Republic of Congo: a new model of rights for forest-dependent communities? In: Helliker K, Murisa T (eds) Land struggles and civil society in Southern Africa, Africa World Press, Trenton.
49. Magnac T (1991) Segmented or competitive labor markets? Econometrica 59:165-187
50. Mookherjee D, Shorrocks A (1982) A decomposition analysis of the trend in UK income inequality. Econ J 92(368):886-902
51. Morris M, Kelly V, Kopicki RJ, Byerlee D (2007) Fertilizer use in African agriculture: lessons learned and good practice guidelines. The World Bank, Washington.
52. Mumvwela CM (2004) Le développement local au Kwango-Kwilu (RD Congo). Peter Lang Publishing, Bern.
53. Ngai R, Pissarides C (2007) Structural change in a multi-sector model of growth. Am Econ Rev 97(1):429-443
54. Nweke FI, Lutete D, Dixon AGO, Ugwu BO, Ajobo O, Kalombo N, Bukaka B (2000) Cassava production and processing in the Democratic Republic of Congo. Collaborative study of cassava in Africa, working paper 22, International Institute of Tropical Agriculture, Ibadan
55. (2012) Agricultural policies for poverty reduction. OECD Publishing, Paris.
56. Otchia C (2013) How could industrial structure guide the choice of development strategy? A field of influence analysis for the Democratic Republic of Congo. Int J Econ Policy Stud 8(1):89-112
57. Otchia C (2013) An integrated approach to trade policy, poverty and income distribution in the Democratic Republic of Congo: a CGE-microsimulation analysis. J Int Dev Stud 22(1):39-53
58. Otchia CS (2014) Distributional and poverty effects of agriculture trade liberalisation: the case of the Democratic Republic of

Congo. Paper prepared for the UNCTAD Virtual Institute project on trade and poverty, UNCTAD-VI, Geneva

59. Ravallion M (1997) Can high inequality developing countries escape absolute poverty? Econ Lett 56:51-57
60. Ravallion M, Chen S (2003) Measuring pro-poor growth. Econ Lett 78(1):93-99
61. Ravallion M, Lokshin M (2008) Winners and losers from trade reform in Morocco. In: Bourguignon F, Bussolo M, Pereira da Silva L (eds) The impact of macroeconomic policies on poverty and income distribution: macro–micro evaluation techniques and tools, The World Bank and Palgrave, Washington. pp 27-60
62. Smale M (1995) "Maize is life": Malawi's delayed Green Revolution. World Dev 23(5):819-831
63. Thirtle C, Townsend R, van Zyl J (1998) Testing the induced innovation hypothesis: an error correction model of South African agriculture. Agric Econ 19(1–2):145-157
64. Thirtle C, Lin L, Piesse J (2003) The impact of research-led agricultural productivity growth on poverty reduction in Africa. Asia Latin Am World Dev 31(12):1959-1976
65. Timmer P (1997) How well do the poor connect to the growth process? Mimeo
66. Vlassenroot K, Huggins C (2005) Land, migration and conflict in eastern DRC. In: Huggins C, Clover J (eds) From the ground up: land rights, conflict and peace in Sub-Saharan Africa, Institute for Security Studies, Pretoria. pp 115-194
67. (2000) World development report: attacking poverty. Oxford University Press, New York.

Chapter 7

Agro-Ecosystem and Socio-Economic Role of Homegarden Agroforestry in Jabithenan District, North-Western Ethiopia: Implication for Climate Change Adaptation

Ewuketu Linger

Debremarkos University, Debremarkos, Ethiopia

ABSTRACT

Homegarden agroforestry is believed to be more diverse and provide multiple services for household than other monocropping system and this is due to the combination of crops, trees and livestock. The aim of this study was to assess socio-economic and agro-ecological role of homegardens in Jabithenan district, North-western Ethiopia. Two sites purposively and two villages randomly from each site were selected. Totally 96 households; in which 48 from homegarden agroforestry user and 48 from non-tree based garden user were selected for this study. Socio-economic data and potential economic and agro-ecosystem role of homegarden agroforestry over non-tree based garden were collected by using semi-structured and structured questionnaires to the households. Homegarden agroforestry significantly ($P<0.05$) improved the farmers cash income than non-tree based garden. With insignificant garden size; homegarden agroforestry practice provides good socio-economical and agro-ecological service for farmers which have a higher implication for climate change adaptation than non-tree based garden.

INTRODUCTION

Background and Justification

Large percentages of the Ethiopian population (80%) depend upon agriculture for their livelihoods, and contribute 42-45% of the total gross domestic product of the country (Zenebe *et al.* 2011). But currently the agricultural production falls under a risk due to a number of factors. Among them; climate change, land degradation in the form of soil erosion, soil fertility loss (which are important for grain yield production) and severe soil moisture stress, which is partly the result of loss of trees in their field and organic matter (Salvatore *et al.* 2011). Regarding climate change the country is one of world's drought prone country, which lead to challenges in food production especially because 95% of the agricultural activity is dependent on rainfall (Eriksen and Kelly2005). Rainfall variability and associated drought have been the major cause of food shortage and famine in Ethiopia (Salvatore *et al.* 2011). The whole effect of the above problem is loss of biodiversity, financial insecurity, food insecurity, subsequent increases in rates of malnutrition, which are becoming the major tribulations of human well-being, so that adaptation to this serious problem should be necessary.

Despite this, one of the solutions to meet diverse people's requirement with fixed land is through the application of agroforestry which is more advantageous than monocropping (Mcneely and Schroth, 2006).

Homegarden with trees are one of agroforestry practices known to be ecologically sustainable and diversifies livelihood of local community. Homegarden is commonly defined as; land use system involving deliberate management of multipurpose trees and shrubs in intimate association with annual and perennial agricultural crops and invariably livestock within the compounds of individual houses, the whole tree-crop, and animal unit is being intensively managed by family labour (Kumar and Nair 2006). Here after, the term homegarden and homegarden agroforestry are used interchangeably throughout this paper.

The high diversity of species in homegarden have a wide socioeconomic and agro-ecological roles including production of food and a wide range of other products such as firewood, fodders, spices, medicinal plants and ornamentals (Unofia *et al.*2012) and avoidance of environmental deterioration of climate related hazards commonly associated with monoculture production systems (Fernandes and Nair 1990), income generating site (Shoo 2009). The diversity of plants in the homegarden associated with other organisms contribute to the formation and maintenance of soil structure, retention of moisture and nutrient levels and promotes the recycling of nutrients; which reduces ecosystem vulnerability to climate change (Verchot *et al.* 2007).

Despite its vast socio-economic importance, in Ethiopia the agro-ecosystem and socio-economic role of homegarden agroforestry are very few.

In order to maintain agro-ecosystem resilience and to meet the homegarden products for requirements of the people during stress of climatic hazard like drought, flood; scientific information is required. Lack of such scientific knowledge of homegarden agroforestry may let destruction of plant diversity and results soil erosion during high rainfall, less income, food insecurity and hunger during drought period. Without a full assessment of socio economic and agro-ecosystem benefit as compared to non-tree based garden, their relation to adapt the changing climate and their role for rural poor people cannot be fully explained. In this particular research non-tree based garden refers to growing of only cereal/cash crop component (maize, teff, barley, wheat, pepper) each of them grown alone or mixed in a household garden rarely with boundary tree component. Therefore this research is aimed at I) Assess the socio-economic role of homegarden agroforestry in comparison to non-tree based garden in Jabithenan district. II) Assess agro-ecosystem role of homegarden agroforestry in comparison to non-tree based garden in Jabithenan district.

MATERIALS AND METHODS

Description of the Study Area

The study was conducted in Jabithenan district, North-western Ethiopia. Geographically the district is found at 10°40′northern latitude and 37°11′eastern longitude.

The topography of the district is generally characterized by flat gentle slope (65%), mountainous (15%), undulating terrain (15%) and valley (5%), with an altitudinal range from 1500 - 2300 m a.s.l. The major soil types found in the district are Vertisol and Nitosol (JWARDO Jabithenan Woreda Agriculture and Rural Development Office 2012).

Climatically, the district falls within midland and lowland agro-ecological zone. The mean annual temperature is about 23°C, with a maximum temperature slightly above 32°C, down to a minimum of 14°C. The mean annual rainfall ranges between 800 – 1250 mm.

The total population of the district is 277,590, of whom 139,616 are males and 137,974 are females. Of this total population, 6% of their population is urban dwellers and the rest 94% are rural dwellers. An estimated population density of the district is about 455.32 people per square kilometers (JWARDO Jabithenan Woreda Agriculture and Rural Development Office 2012).

Historically, the district was covered by dense natural forests, but the distribution of natural forest is declining from time to time, owing to human interference. The common vegetation in the district include, *Croton macrostachys, Ficus sur, Albizia gummifera, Cordia africana, Acacia abyssinica, Rosa abyssinica* and *Erythrina abyssinica* which are found as scattered in most farm lands. While*Eucalyptus* spp and

Gravillea robusta are grown as boundaries, live fences and woodlots (JWARDO Jabithenan Woreda Agriculture and Rural Development Office 2012).

Agriculture is the principal source of livelihood for rural population. It is characterized by subsistence mixed farming of rain-fed, irrigated crops, and livestock production together with trees planted as an agroforestry. In the district cereal crops are the staple food crops such as teff (*Eragrostic teff*), maize (*Zea mays*), wheat (*Triticum sativum)* and barley *(Hordeum vulgare)* are the most commonly cultivated crops. Pepper (*Capsicum frutescens)*, sugarcane (*Saccharum officinarum)*, coffee (*Coffea arabica*) and chat (*Catha edulis*) are the dominant cash crops in some sites including the study village (JWARDO Jabithenan Woreda Agriculture and Rural Development Office 2012).

Sampling Method

Purposive sampling method was employed. Two sites (here after Kebele Administrations[a], KA) namely, Mankusa Abdegoma and Jiga Yelmdar were selected purposefully based on the extensive presence of both homegarden agroforestry (HGAF) and non-tree based garden (NTBG). Potential villages (five villages from Mankusa Abdegoma and three villages from Jiga Yelmdar Keble) were identified by reconnaissance survey with district and Kebele agricultural office experts. Then Debohela and Waza villages from Mankusa Abdegoma KA; Atahagne and Tikurwuha villages from Jiga Yelmdar KA were randomly selected.

[a]KA is the lowest administrative unit in Ethiopian government structure

In this study, key informants and households were involved to assess how HGAF is important to them over NTBG. KIs for this study is defined as persons who are knowledgeable about the previous and current situation of local climate conditions and role of various HGAF for climate change adaptation, agro-ecosystem maintenances and who lived there at least for continuous 30 years. KIs were selected by snowball method (Bernard 2002). Accordingly during village reconnaissance, five farmers were randomly asked to give the name of six KIs. Out of the mentioned thirty candidate KIs, the five top ranking were selected at each village which make a total of 20 KIs for the entire study. Those selected KIs with DA were also used to classify household into two garden types and wealth categories based on local wealth criteria. HH is defined as a basic unit of production and consumption, made up from the persons who manage common landholdings and live less than one central decision-maker, the household head. Finally, from each wealth class 8 HHs were picked randomly (4 HGAF and 4 NTBG practicing HH), making 24 HHs per village and 96 HHs for the entire study.

Data Collection

Based on information from KI interview, questionnaires were designed to collect data on the role of gardens in socio-economic and agro-ecosystem maintenance and associated implication when climate variability/change occurs was collected. Focused group discussions were also used to validate the information given by an individual farmer and to catch important issues that were not raised by respondent farmers.

Two years (2011/2012 and 2012/2013) HH income from two garden types was collected. Product from HGAF that were consumed by household as food were not converted in to cash, since in every time children's are using especially fruits and tubers of herbs for food and farmers are unable to determine the exact value. Crops produced from NTBG were converted in to cash according to the current market during data collection time. Both socio-economic and inventory data were collected from early November to end of December 2012 while garden income of the year 2012/13 were collected in one month (April 2013).

Data Analysis

Data collected from the interview and questionnaire was coded, enters in to a computer, analyzed, interpreted and synthesized using Statistical Package for Social Sciences (SPSS) software Version 16. Socio-economic and agro-ecosystem role between garden types was subjected to one-way ANOVA. Mean differences between groups were considered significant at $p < 0.05$.

RESULTS

General Evidence of Climate Variability in the Study Village

Particularly rainfall and temperature for the study kebeles were taken from metreology agency of the nearby station (Figures 1 and 2). For Mankusa Abdegoma Kebele[b] lay-ber station was the source which is 10 km far from the study Kebele. Whereas for Jiga Yelmdar, Yechereka station was used having 6 km distance (NMA National Metreology Agency 2012). MAT[c] and MARF[d] were used.

[b]Kebele is the lowest administrative unit in Ethiopia.

[c]MAT, Mean annual temperature (mean of 12 month temperature in a year).

[d]MARF, Mean annual rain fall (mean of 12 month rainfall in a year).

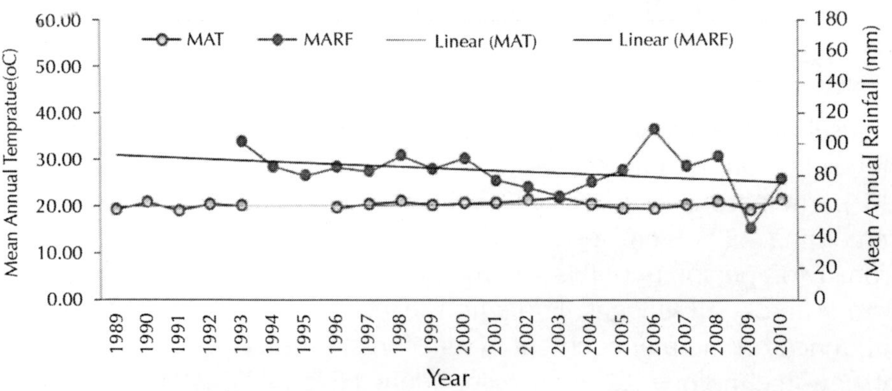

Figure 1: Mean Annual rainfall and temperature from 1989-2010 in Mankusa Abdegoma Kebele, Jabithenan district, Ethiopia.

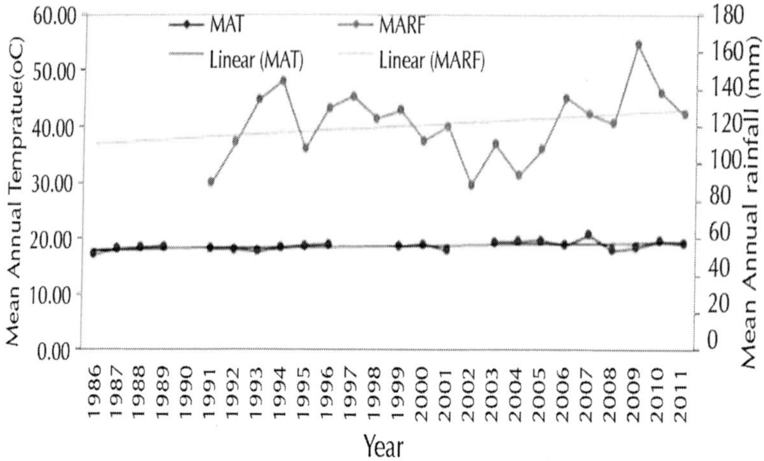

Figure 2: Mean annual rainfall and temperature from 1986-2011 in Jiga Yelmdar Kebele, Jabithenan district, Ethiopia.

Socio-Economic Characteristics of Respondents

Household's socioeconomic features can influence on-farm/garden tree selection, management and species diversity. In this study, of the 96 participating households, 78% were male headed whereas the remaining 22% were female headed. Number of dependent ranged from 1 to7 person per HHs and the number of labor force ranged from 2 to 9 across all villages. Education levels of a HH have also a direct influence on the management of agroforestry practices or in adoption of new technology. Of total respondent HHs 64%, 21%, 10%, and 5% were illiterate; attended Grade 1-4, 5-8 and secondary school respectively.

General Features of Garden Types

Homegarden Agroforestry (HGAF)

The original layout of most gardens was established during the villagization program of the military (Derg) regime (1974-1992). Then after producing cereal crops for some years, in 1992 about19%, in1995 (67%), and 14% of households in 1996 were integrate perennial trees in their garden and becomes HGAF (n=48).

Spatial arrangement of HGAF is variable at the study site. About 74% of HGAF is located in the backyard while a few are located on the side of the homestead. Most of the studied HGAF (67%; n=48) are surrounded by live fence of the species *Eucalyptus camaldulenesis, Capparis tomentosa, Rosa abyssinica, Carissa edulis, Combretum molle, Maesa lanceolata, Bersama abyssinica, Acacia nilotica*. The rest 33% are semi-fenced and open.

The overall HGAF size ranged from 0.05 to 0.5 ha (0.1 to 0.5 ha in Jiga Yelmdar and 0.05 to 0.38 ha in Mankusa Abdegoma Kebele). The average size of HGAF in Jiga Yelmdar and Mankusa Abdegoma Kebele is 0.21 ha and 0.16 ha respectively. Most of the surveyed HGAF (63%) in Mankusa Abdegoma and (60%) in Atahagne village of Jiga Yelmdar kebeles have rectangular shape while the remaining were irregular and square shape.

Non-Tree Based Garden (NTBG)

Like HGAF, the overall NTBG size ranges from 0.05 to 0.5 ha. But in contrary, NTBG were covered by cereal/cash crop (Figures 3 and 4). From the 48 surveyed NTBG, 7 numbers of species (i.e. species richness) were recorded. Those species are Maize (*Zea mays*), Potato (*Solanum tuberosum)*, Wheat (*Triticum sativum*), Barely (*Hordeum vulgare*), Sugarcane (*Saccharum officinarum)*, and Onion (*Allium cepa)* and pepper (*Capsicum frutescens*). Figures 3 and 4 indicates the frequency of cereal crop recoded out of the 48 interview respondents households in two successive years at each village. Onion, pepper and sugarcane are an important source of cash for the household.

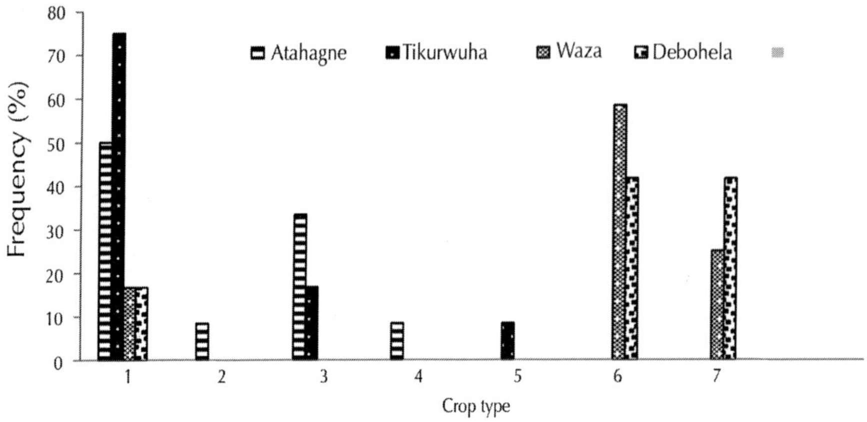

Figure 3: Types of crops produced in non-tree based garden in year 2012 and proportion of respondent households at four villages, Jabithenan district, Ethiopia.

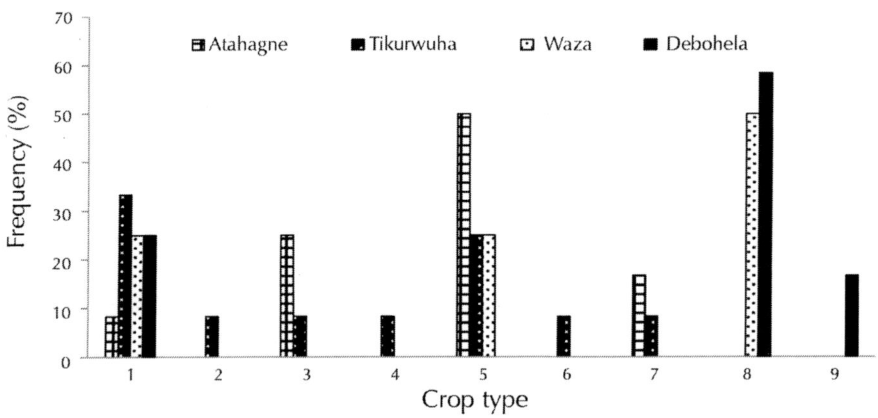

Figure 4: Types of crops produced in non-tree based garden in year 2013 and proportion of respondent households at four villages, Jabithenan district, Ethiopia.

Advantage of Homegarden Agroforestry Over Non-Tree Based Garden

HGAF plays an important role and unique land management system because of the potential role in addressing biophysical, economical and socio-ecological components. Such diversity and interaction leads to greater functional and structural complexity as compared to non-tree based garden. Surveyed households admire homegarden importance especially during climate related shock happen. Farmers grouped climate related hazards in to three major topics: climate variability (100), drought (87), rain with ice (92) which hampers production of cereal crops. The value in the bracket shows respondents percentage.

Socio-Economic Role

Economic difference was great among the two garden types with almost proportional social group (religion, age group, work habit, land size, homegarden size), infrastructure and extension service. The focused group discussions employed also indicated that production of cereals in NTBG is limited to only rainy season (Table 1).

Table 1: Summarized characteristics of HGAF and NTBG at four villages, Jabithenan district, Ethiopia

	Garden Type	
	HGAF	NTBG
Major characteristics and use for the household	High diverse species	Less diverse species
	Production is throughout the year	Production is once, rarely 2 times (irrigation)
	Saves working time during peak labor work month through provision of immediate fresh food	Households wait until the food is cook, time wasted
	High labor cost	Less labor cost
	Reduce cost of purchasing fertilizer	Use chemical fertilizers (Incurs cost)
	Increase social intimacy through coffee ceremony	Involvement in coffee ceremony is limited to relatively well-off HHs.
	Use of fruits reduce frequency of request for meals and avoid hunger especially for children	Relatively higher frequency of request for meals 3 to 5 times/day

The study revealed that, different species grown in the homegarden gives different functional roles for the household. In HGAF, 77 and 67% of respondent consume fruit trees as food and generate cash income respectively. HGAF also provide fuel wood much greater than NTBG, which is the major energy source in the site. In addition to home use, some households get an income from fuel wood selling and 19% of respondent farmers also prepare charcoal from *Acacia abyssinica*.

Linger

Linger *SpringerPlus* 2014 3:154 doi:10.1186/2193-1801-3-154

Households in the study area generate income from different homegarden products. Income results from two garden type shows significant difference (Table 2).

Table 2: Mean farm size, garden size, annual income of 2011/12 and 2012/13 from the two land use types of four villages, Jabithenan district, Ethiopia

Village	Garden type	Farm size(ha)	Garden size(ha)	Cash income (Birr) 1$ = 19birr when the research conducts	
				Year 2011/12	Year 2012/13
Tikurwuha	NTBG	1.67a ± 0.16	0.25a ± 0.03	2,266.67aA ±349.31	1,612.50aA ± 351.73
	HGAF	1.69a ± 0.18	0.22a ± 0.02	15,750.00bB ± 3735.83	10,050.0bB ± 3345.54
Atahagne	NTBG	1.77b ± 0.20	0.17b ± 0.03	1,867.33cC ± 308.06	1,353cC ±210.27
	HGAF	2.15b ± 0.24	0.23b ± 0.03	22,483.33dD ±9361.23	13,112dD ± 4351.57
Debohela	NTBG	1.17 c ± 0.09	0.14c ± 0.02	5,825.00eE ±1354.46	2,487.5eE ± 905.71
	HGAF	1.13c ± 0.10	0.10c ± 0.01	6,858.33eF ± 1549	5,857.5eF ± 1914.10
Waza	NTBG	1.00d ± 0.09	0.18d ± 0.03	5,987.50fG ±1716.81	770.00fH ± 149.21
	HGAF	10.21d ± 0.10	0.14d ± 0.02	6,945.83fI ± 1609.72	7,033.33gI ± 2298.40
Overall mean	NTBG	1.40 ± 0.14	0.19 ± 0.03	3,986.50 ± 932.16	1,555.63 ± 404.23
	HGAF	1.55 ± 0.16	0.17 ± 0.02	13,009 ± 4063.95	9,013.21 ± 2977.40

NB - Single different small letters on mean values indicate significant difference at ($P < 0.05$) between the two land users with in the village in a column.

Single different capital letters on mean values indicate significant difference of income at ($P < 0.05$) between the two years with in the same land use in the village horizontally (row).

Linger

Linger *SpringerPlus* 2014 3:154 doi:10.1186/2193-1801-3-154

Agro-Ecosystem Role

HGAF play an important role for agro-ecosystem service mainly through providing raw material for compost production (Table 3). Without significant difference in the demographic character and land or homegarden size between garden types (Table 2), 100% of the respondent confirms that the different in the volume of compost produced was due to the raw material (weeds and grasses, tree leafs) available within a HGAF. Using HGAF for compost production 94% of the respondents (n=48) say that, the fertility status of the soil stays up to a minimum of three years and a maximum of four years in addition to soil moisture conservation while rainfall shortage happens (Table 4).

Table 3: Mean volume of compost and number of livestock/household in the two garden types at four villages, Jabithenan district, Ethiopia

Village	Garden type	Volume of compost (m3)	Number of livestock
Atahagne	HGAF	23.67a±4.18	14a±1.94
	NTBG	9.92b±0.99	7b±0.84
Tikurwuha	HGAF	25.5a±3.01	16a±2.34
	NTBG	9.08b±1.27	6b±0.84
Waza	HGAF	16.58a±3.29	7a±1.14
	NTBG	9.33a±2.17	12a±1.79
Debohela	HGAF	21.58a±2.97	7a±1.40
	NTBG	7.64b±2.13	5a±0.47
Overall mean	HGAF	21.83±3.37	11±1.70
	NTBG	8.99±1.64	8±0.99

NB - Single different letters on mean values indicate significant difference at ($P<0.05$) between garden types in a village in a column.

Linger

Linger *SpringerPlus* 2014 **3**:154 doi:10.1186/2193-1801-3-154

Table 4: Percentage response of two garden type users' on gardens ecosystem role at four villages, Jabithenan district, Ethiopia

Parameters	Percent response	
	HGAF (n=48)	NTBG (n=48)
Inorganic fertilizer application in garden	0	81
Herbicide/pesticide application in the garden	0	87
Presence of erosion in garden	0	92
Use of animal dung for soil fertility	77	12
Crop residual removal from outside garden	40	92
Presence of moisture during rainfall shortage	94	15
Garden fodder tree increase livestock number	71	0
Compost reduce fertilizer coast	85	35

Linger

Linger *SpringerPlus* 2014 3:154 doi:10.1186/2193-1801-3-154

HGAF provide fuel wood source which leads to less farm crop residual biomass removal, less dependence on animal dung for fuel and minimum investment of money for organic fertilizer than NTBG (Table 4).

DISCUSSION

Homegarden Agroforestry: Contribution and Implication to Climate Variability/Change Adaptation

The species in the NTBG was composed of mainly cereals and their coverage/type of species varies in time/year. In 2011/2012 onion was not recorded; but was planted in year 2012/2013 during winter through irrigation. The reason is because as 100% of respondent confirms, there was seasonal (three weeks) rainfall shortage (drought), crops were failed due to less water availability during growing season in 2012/13. Even in the beginning, onset of rain was late (June 22, 2012) and lags from the normal rainfall year (early may), which hampers crop production. Then NTBG users immediately react/cope by doing income generation activity basically onion irrigation from the distance water, which is labor intensive. Since HGAF has diverse component than NTBG, risk was lower in HGAF and have the following role.

Food and Income Role

Species in the studied HGAF supplement for household's food and income. Tubers and vegetables are among the herb species and replace the staple food of cereal crop. Research results also showed that tubers are replacing the basic staples of cereals, and produce reasonable amount of carbohydrate (Fernandes and Nair 1986). Respondents assure that fruit trees are primarily served for food especially during difficult time of drought. The three fruit tree species frequently found was an indicator of how farmers are highly depends in HGAF for food in addition to their cash income revenue. Relatively *Persea americana* and *Mangifera indica* have a good market (but not consistent) in the near zonal town, Fintoselam than other fruit tree species. This is due to the fact that urban peoples use the processed juice from the two fruit tree species. Fentahun Mengistu (2008) also confirms, fruit trees from HGAF have significant role during environmental crisis of households. While family/especially children's consume fruit trees, it is possible to avoid (buffer) frequency of hanger and decreases number of meals/day. So dependency on other cereal food crops becomes minimal. In the long run, cereal crops are reserved/saved to buffer households during time of stress. In contrary, in NTBG land user this is not possible since household members have no other substitution food items to save cereal crops for difficult time, and so the vulnerability becomes higher than HGAF user. Research findings from South Africa agroforestry fruit trees show, fruits play an important role especially during time of famine and other stress as food, nutrition and cash income (Akinnifesi *et al.*2008). All other components of homegarden in the study area help the same as what fruit tree does in one way or other. According to Scherr (1995), rural households survival strategies encompass multiple objectives in maximization of utility, like provision of food, and subsistence goods, cash for purchase of goods, services and saving for future needs, while environmental or other shock happens; in which HGAF can do. A study in Zambia showed that some products from homegardens, particularly fruit and cabbage are important buffers during drought years (Alfred 2009) and consistent with this study.

Income result between the garden types in Jiga Yelmdar was significant, unlike Mankusa Abdegoma Kebele in 2011/2012. Sugar cane and pepper both a high value marketable crop were dominantly occurred in Mankusa Abdegoma.

In 2012/2013 income result from Jiga Yelmdar kebeles was significantly higher in HGAF than NTBG. This explanation could be due to the double factors (drought and less value market crops in NTBG).

In 2012/2013 income results from Waza village of Mankusa Abdegoma kebeles was significantly ($P<0.05$) higher in HGAF. Even NTBG were covered by high marketable crops, insufficient moisture content contribute to the reduction in cash income. In this village the income was not significantly vary in 2011/2012, since no drought year. A study in Amhara region, Ethiopia showed that production of cereal crops (wheat, barely, maize, teff, sorghum) showed statistically significant correlation with seasonal rainfall variability during 1994-2003 (Bewuket 2009).

Current research Journals confirmed that, relaying in HGAF remains viable for foreseeable future as a strategy to stabilize household food security and income against the risks uncertainties of monocropping due to multiple products (Alfred 2009). Other findings also confirm that HGAF contributes to household risk management via reducing income variability (FAO Food and agricultural organization 2011).

The sale of products from homegarden significantly improves family financial status, and cash income can be used to buy food, cloth etc. This observation is in line with the study done in homegarden of Zimbabwe and Ethiopia (Alfred 2009; Kebebe and Urgessa 2011).

The presence of fodder tree has also a crucial value, increases the number of livestock (important assets during time of crisis). Review report from dry land of Africa showed that presence of fodder tree in the garden serves not only increase number of livestock but also reduces livestock forage cost (Bashir *et al.* 2006). Other findings also show that an increase in number of livestock leads to an increase in food stock due to livestock waste is useful in the farming to improve crop yield (Anne 2008).

Ecological Role

The multi-layered, forest like vegetation structure of the studied homegarden in the area contributes substantially to the agro-ecological sustainability through reducing soil erosion. Research findings from homegarden of Meghalaya, North-east India also confirms that, multilayered vegetation structure prevents soil erosion, provides habitat to soil micro-organisms and promote a favorable microclimate for the household (Tynsong and Tiwari 2010). Fernandes and Nair (1986) also reported that the presence of multi-layered structure of homegarden is an indicator of ecological function through environmental protection and efficient use of resource (like sunlight).

HGAF play a role not only erosion control in the garden, but also give three or four year soil fertility maintenance out of the garden/other farm field and saves soil moisture through composting. Compost is not only give soil fertility and moisture conservation, but also reduces the farmer's fertilizer cost (could be invested in other economic activity like rent land to produce another crop). Alfred 2009 cited in Chivaura-Mususa *et al.* (2000) showed that livestock manure, litter fall and compost was found a rich in nitrogen and becomes a good fertilizer to maintain soil fertility and crop production. The result obtained here is also in line with other findings in homegardens of Wayanad, Kerala; homegarden supply raw materials (such as leaf and compost) to agricultural land for the local community (Santhoshkumar and Lchikawa 2010). Bishaw *et al.*(2013) reported that the higher number of livestock due to fodder tree availability provide significant amount of manure that can go in to improving soil fertility. HGAF give fuel wood; reduce pressure on animal dung. So farmers of HGAF user were applying this dung for soil fertility. Whereas majority of NTBG, animal dung were used for both fuel wood and soil fertility; and purely for fuel wood.

HGAF also helps for crop residuals to be remaining there in the farm (since HGAF serves what crop residual serves) after the grain harvest. Residuals in the process decompose and help for soil fertility. In NTBG majority of the respondent harvests residuals for livestock feed and fuel wood, which reduce soil fertility. Therefore, the overall soil fertility maintenance, soil moisture conservation help the crops to be grown, not to wilt when rain fall variability happens for one/two weeks.

Social Role

Some households gave some homegarden products (like fruits, vegetables, sugarcane) to neighbors and relatives, which strengths nebiour and family relationship called social capital. When there is a religious day (e.g. Saint Gebreal and Micael), rural peoples have a coffee ceremony in circulation with their neibour. But a person/family who has not coffee in his garden was not participating in the ceremony, since he/she can't able to return in other days. NTBG users participate, if they are able to purchase coffee from the market. In this case coffee as a HGAF component increase social relationship and sharing different working experience (regarding homegarden species management, harvest) from model farmers, unlike NTBG users. This observation is in consistent with the results from research in Zambia, homegarden of Meitei community of India (Alfred 2009; Davi and Das 2010) helps for good social relation through sharing of homegarden vegetables. The overall implication is increasing intimacy and farmers can help each other when a farmer/household becomes food insecure. Researchers from social science agree that, social capital reduces/adapt risks of climate change; avoid conflict through interaction between people or state (Neil 2003; Bishaw *et al.* 2013).

CONCLUSION AND RECOMMENDATION

The associated component diversity of Homegarden agroforestry enhances the livelihood of the local people by providing socio-economic and agro-ecological service than non-tree based garden; important to adapt climate change or climate related stress (drought).

The government should work jointly with the local farmers in changing monocrop land use system to mixed agroforestry system for climate change adaptation.

ACKNOWLEDGEMENTS

I thank the Ethiopian ministry of education and Debremarkos university for financial support of this research which otherwise would face financial constraints. I thank Jabithenan Woreda Agriculture and Rural Development Office, for providing me secondary data sources and field equipment's for the field work activities. My special thanks to key informants and farm households in each Kebele for their kind respect and allowing me to collect data both from them and their Homegardens.

REFERENCES

1. Akinnifesi FK, Sileshi G, Ajayi OC, Chirwa PW, Kwesiga F, Harawa R (2008) Contribution of agroforestry research and development to livelihood of smallholder farmers in South Africa. Agric J 3:76-88
2. Alfred M (2009) Traditional homegardens and rural livelihood in Nhema, Zimbabwe: a sustainable agroforestry system. Int J Sustain Dev World Ecol 16:1-8
3. Anne A (2008) The role of home gardening in household food security in Butere Division of the western Kenya. Phd dissertation.
4. Bashir J, Eyasu E, Kebadire M (2006) Role of agroforestry in improving food security and natural resource management in dry lands: a regional overview. J Dry Lands 1:206-211
5. Bernard HR (2002) Research Methods in Anthropology: Qualitative and Quantitative Approaches. Walunt creek, CA: Altamira.
6. Bewuket W (2009) Proceeding of the 16th International Conference of Ethiopian Studies. Trondheim, Norway.
7. Bishaw B, Henry N, Jermias M, Abdu A, Jonathan M, Gemedo D, Tewodros A, Kathleen G, Habtemariam K, Ian K, Eike L, Cheikh M (2013) Farmer's strategies for adapting to and mitigating climate variability and change through Agroforestry in Ethiopia and Kenya. Corvallis, Oregon: Oregon State University.

8. Chivaura-Mususa C, Campbell B, Kenyon W (2000) The value of mature trees in arable fields in the smallholder sector, Zimbabwe. Ecol Econ 33:395-400
9. Davi N, Das K (2010) Plant species diversity in the traditional homegardens of Meitei community: a case study from Barak Valley, Assam, India. J Trop Agric 48:45-48
10. Eriksen K, Kelly PM (2005) The dynamics of vulnerability: locating coping strategies in Kenya and Tanzania. Geogr J 171:287-305
11. FAO (Food and agricultural organization) (2011) Climate-Smart Agriculture: Small Holder Adoption and Implication for Climate Change Adaptation and Mitigation. Rome Italy.
12. Fentahun Mengistu (2008) Fruit Tree Species in the Wild and in Homegarden Agroforestry: Species Composition, Diversity and Utilization in Western Amhara Region, Ethiopia. PhD. Thesis. Vienna: Vienna University.
13. Fernandes ECM, Nair PKR (1986) An Evolution of the structure and function of tropical homegardens. Agr Syst 21:279-310
14. Fernandes ECM, Nair PKR (1990) An Evaluation of the Structure and Function of Tropical Homegardens. In: Landauer K, Brazil M (eds) Tropical Home Gardens, Tokyo: United Nations University Press. pp 105-114
15. JWARDO (Jabithenan Woreda Agriculture and Rural Development Office) (2012) Jabithenan Woreda Agriculture and Rural Development Office, the Annual Report. Jabithenan, Ethiopia.
16. Kebebe Z, Urgessa K (2011) Agroforestry perspective in land use pattern and farmers coping strategy: Experience from southwestern Ethiopia. World J Agric Sci 7:73-77
17. Kumar BM, Nair PKR (2006) Tropical Home Gardens: A Time-Tested Example of Sustainable Agroforestry. India: Environmental Experts.
18. Mcneely AJ, Schroth G (2006) Agroforestry and biodiversity conservation traditional practices, present dynamics and lessons for the future. Biodivers Conserv 15:549-554
19. Neil A (2003) Social capital, Collective action, and adaptation to climate change. J Econ Geogr 79:387-404
20. NMA (National Metreology Agency) (2012) National Metreology Agency of Ethiopia. Ethiopia: Bahirdar branch.

21. Salvatore Y, Gunnar K, Ringler C (2011) Estimating the impact of climate change on agriculture in low income countries: Household level evidence from the Nile basin Ethiopia. J Environ Res Econ 28:32-43
22. SanthoshkumarV, Lchikawa K (2010) Sustainable use of Biological Biodiversity in Socio-Ecological Production Landscapes in Wayanad, Kerala. Info Change News and Features. India: United nations University institute of Advanced studies.
23. Scherr (1995) Economic Analysis of Agroforestry Systems: The Farmer's Perspective. Washington, USA: World Bank project.
24. Shoo UK (2009) Traditional homegardens and livelihood security in north-east India. J Food Agric Environ 7:665-670
25. Tynsong H, Tiwari K (2010) Plant diversity in homegardens and their significance in the Livelihoods of War Khasi community of Meghalaya. North East India J Biodivers 1:1-11
26. Unofia SI, Owoh PW, Ukpong EE, Ekpo IE (2012) Assessment of plant species of socio-economic importance conserved in homegarden of Nsit ubium local government area of Akwa Ibom state, Nigeria. Nig J Agric Food Environ 8:99-108
27. Verchot LV, Van Noordwijk M, Kandji S, Tomich T, Ong C, Albrecht A, Mackensen J, Bantilan C, Anupama KV, Palm C (2007) Climate change: linking adaptation and mitigation through agroforestry. J Mitig Adapt Strat Glob Change 12:901-918
28. Zenebe G, Jesper S, Alemu M, Atlaw A (2011) Climate Change and the Ethiopian Economy. A Computable General Equilibrium Analysiss. Ethiopia: Environment for Development.

Chapter 8

Identifying and Measuring Land-use and Proximity Conflicts: Methods and Identification

André Torre[1], Romain Melot[1], Habibullah Magsi[2], Luc Bossuet[1], Anne Cadoret[3],Armelle Caron[4], Ségolène Darly[5], Philippe Jeanneaux[4], Thierry Kirat[6], Haï Vu Pham[7], and Orestes Kolokouris[8]

[1]UMR SAD-APT, INRA AgroParisTech, Paris Saclay University, Paris, France

[2]Department of Economics, Sindh Agriculture University, Sindh, Pakistan

[3]UMR Telemme, Aix-Marseille University, Aix-Marseille, France

[4]UMR Metafort, AgroParisTech, Clermont Ferrand, France

[5]University Paris VIII Vincennes Saint-Denis, Paris, France

[6]CNRS, University Dauphine, IRISSO, Dauphine, France

[7]UMR CESAER, INRA AgroSup, Dijon, France

[8]Panteion University, Athens, Greece

ABSTRACT

This text aims to present the methodology of study of land-use conflicts performed in recent years by a multidisciplinary team, and to reveal the methods of survey and data collection, as well as the structure of the resulting database. We first define the scope of our study by providing a definition of these conflicts, of their characteristics and motives, of the ways they manifest themselves and of the actors involved (I). We then present the methodology we have used to identify conflicts; it is based on a spatial analysis and the combined use of different data collection methods including surveys conducted by experts, analyses of the regional daily press and of data from the administrative litigation courts (II). Finally we present the resulting *Conflicts* © data base, with its tables and nomenclatures, in which the data collected in different fields are reconciled and analyzed (III), before providing a few examples of how this method can be used to analyze case studies in developed and developing countries (IV).

INTRODUCTION

Though conflict analysis is inscribed in a long tradition of social sciences, at the first rank of which lies sociology (Lewin, 1948; Touraine, 1978; Stephenson, 1981; Simmel, 2008; Freund, 1983; Coser, 1982; Wieviorka, 2005), researchers and practioners have preferred to focus their attention on the questions of conflict resolution rather than on the analysis of conflicts and of their characteristics (Castro and Nielsen 2001; Jeong 1999; Fisher, 1997; Neslund, 1990; Owen et al., 2000), except in cases of armed conflict (Boulding, 1962; Diehl, 1991; Hensel, 2001; Starr, 2005). Yet, the growing concerns about the environment, the issues of sustainable development, urban sprawl processes and about questions about people's living environment have recently led to renewed interest for issues related to land use conflict, also called land use and neighbourhood conflicts or environmental conflicts (See, among many others: Humphreys,2005; Deininger and Castagnini, 2006; Magsi and Torre 2014; Mann and Jeanneaux, 2009; Campbell, et al., 2000; Darly and Torre 2013a, b; Cadoret 2009; Melé et al. 2004; Dziedzicki2001; Charlier 1999; Cadene, 1990).

Interest in these issues has grown in the fields of economics, geography, land planning as well as in sociology and social-psychology and has pointed to the necessity of analyzing conflicts, their occurrences, their impacts and main characteristics, more thoroughly. Such an approach obviously necessitates access to enough reliable data on conflicts per se, so as to be able to evaluate their number and volume, their role and impact, how they manifest themselves, what their causes or origins are and how they are solved.

Data about conflictuality is scarce and often incomplete for two main reasons. The first lies in the fact that little interest was shown in the subject until the years 2000. The second reason is related to the complexity of conflicts, which rules out the possibility of using only one representative variable. Indeed, land use conflicts find expression in various forms (tribunals, media coverage, violence...) which prevent one from making a simple representation and which explain why various disciplines are necessary to define them. A conflict that gives rise to analysis is a construct founded on information collected from different sources.

In light of this problem, a researcher wishing to study conflictuality must collect his/her own data (either they are focusing on developed or developing countries), and then analyze them (for more on the subject read Rucht and Neidhardt (1999) which describes the different stages that are necessary to this type of work). This is what we have done by developing a program of study on conflicts involving several teams of French researchers from various research institutes and universities. This program has successively focused on various issues pertaining of conflicts in natural, rural and peri-urban areas in developed and developing countries (France, Greece, Canada and Pakistan). Their contributions are based on analytical method adopted according to the nature and availability of data from different economies, and identification of land use conflicts, involvement of actors with different relations and network links.

This text aims to present the methodology of study of land-use conflicts performed in recent years by a multidisciplinary team, and to reveal the methods of survey and data collection, as well as the structure of the resulting database. The originality of this approach lies in its refusing to only use one particular source of information or one single innovative formula. The method we use to identify the conflicts, and which we present here, is complex and multi-dimensional. It involves a specific methodology that rests on the combination and triangulation of different sources and modes of data collection, and on precise protocols of data processing and of identification of conflict patterns, applied in each stage of the program. Following these protocols ensures that we achieve the most realistic representation possible of conflict within a given space or area. Our method borrows from other investigation procedures previously developed (See for example, Charlier (1999) based on press articles) and which we have transposed, improved and specified, but it also rests on innovative procedures (the use of litigation data and their treatment by means of software). It is based on social science analysis techniques (statistical inquiries, interviews, surveys, accounts, group follow ups…). It also consists in exploiting databases (such as the *Lamyline* database) or data provided by administrations (such as tribunals' judgments).

In some ways, our approach is not unlike that of other current projects of research on land use conflicts, in particular urban ones (see the research conducted in Canada (Trudelle, 2003; Joerin et al., 2005), in Netherlands (Leeuwen, 2010), or in Brazil (Observatorio Permanente dos Conflitos Urbanos na Cidade de Rio de Janeiro 2010). It is situated within a tradition that includes remarkable works conducted by teams of researchers located in various countries. Among these studies, three are, in our view, particularly significant, and have paved the way for our work by indicating the main difficulties involved in this type of research and by suggesting many solutions and possible paths for further research (Janelle, 1977; Ley and Mercer, 1980; Rucht and Neidhardt, 1999). Although we have not always drawn the same conclusions or adopted all the suggestions made by those authors, their works have been essential sources of inspiration. We owe a great deal to these studies in that they have helped us reflect on our multidimensional method of conflict analysis and avoid many pitfalls.

Two reasons explain why we have wished to present our method of analysis – which we have tested and improved over the last 7 years – to a wide audience of researchers and practitioners:

- We have wanted to show that it is possible to identify conflictual events and to deduce from them a general picture of conflictuality as well as a description of the characteristics of the conflicts that occur in a given area;

- We have also wanted to share our experience and encourage researchers to use our method, by responding to its associated criteria of intellectual property.

The structure of this article is as follows: we shall first present the scope of our investigations, placing particular emphasis on the questions of definition, origins and expression of conflicts. The second section of the paper is devoted to the presentation of our data collection method. We begin by describing the procedure of identification and diagnosis of the conflicts that occur in the area we have selected, and then discuss the identification method per se and more specifically its three foundation stones: the analysis of the daily regional press, of administrative litigations and the exploitation of surveys conducted by experts. In the third section we first present the *Conflict* © database compiled from the information gathered. It addresses the questions of the construction and classification of the objects of conflicts, of the actors› profiles, of their utilization of land and of their arguments, successively. We end the article with a brief presentation of the results obtained from the exploitation of the database on conflictuality on selected case studies from developed to developing countries.

Definition of the Scope of Investigations

We are interested in land use conflicts and conflicts over resources. In order to identify them, it is, first, necessary to propose an operational definition that will enable us to both recognize and capture conflictual elements and situations, and to classify them in such a way as to be able to trace the conflictual profiles of a given area.

Our research studies are conducted in rural and peri-urban territories. They pertain to conflicts and tensions related to public consumption commodities (air, landscape amenities and nature's functions), resources (water or energy), waste and pollution, as well as to the areas of location of individuals or activities and their neighboring areas.

We call conflict an opposition marked by an engagement or a commitment between two or several parties (the actors of the conflict), in relation to local material objects. These oppositions reveal local characteristics related to spatial dimensions (e.g. topologic dimensions, neighborhood and transport infrastructures) as well as to social and economic characteristics linked to the areas in which they arise. Land use conflicts are the results of the dissatisfaction of one part of the population with actions undertaken or planned by their neighbors, by private institutions or by the public authorities. They are the pointers of the innovations taking place in the different territories and of the resistance they generate, and also the ferment of new innovation phases. We do not consider it necessary to eliminate conflicts nor even to try and solve them at all costs, for they are the expression of the voiced opposition of parties that consider themselves injured. Conflictual events are phases of coordination between actors and a way of reintroducing new actors in the mechanisms of decision making and of creation of territorial development projects. Conflicts can often be considered as part of a trial and error method regarding public decision: if one decision is considered opposite to the needs and wills of local populations, it could lead to tension, and afterwards to conflicts.

A Local Materiality

The conflicts we are interested in are distinguishable by their localized nature (i.e. territorial superposition of contradictory interests, rivalries between contiguous or neighboring areas), by the materiality of the objects that cause them or are concerned by them, as well as by the fact that they emerge in relation to differing land uses. Oppositions between individuals or groups pertain to concrete objects, to technical acts that are taking or will take place and imply concrete actions. These conflicts can have a strictly local component, or be related to questions that are more universal in scope. Whatever the initial situation, they can expand socially and spatially if they crystallize issues of a societal nature.

Land use conflicts have a territorial dimension. They rest on a physical basis: they take place between actors (sometimes, but not always neighbors) affected by a problem that has emerged and they develop around the use of localized support material or immaterial goods. They are inscribed in a geographic institutional framework, determined by both the actions of local and supra local authorities and by the rules they introduce. Indeed, *territoriality* and *unequal exposure* are central characteristics of land use conflicts. Environmental disputes are, first of all,*territorial* in nature to the extent that they involve the very unequal exposure of different territorial regions to environmental pollution, land speculation and development projects. Regions where the highest conflicts concentration is observed are also the regions where the densest concentration of risk-bearing facilities or building permits can be found.

Conflictual events are identifiable in relation to specific goods or pieces of land, i.e. the space within which the uses are in opposition. The cases studied in our research pertain to questions related to land, to territorial development as well as to water and its management, to the superposition of uses (production, tourism, leisure), to the development of economic industrial and port operation activities, to landscape and their evolution through urbanization and the creation of new equipment such as wind turbines, waste water treatment plants, waste management facilities, etc. More specifically, researches implemented in a country like France have been focused on conflicts dealing with the issue of urban sprawl and its consequences on speculation in farm land and forests (in particular risk of arson in peri-urban forests). In a Mediterranean country like Greece, forest fires can be mobilized as a strategy of land appropriation. Arsons can be used in speculative strategies to change land use destination, but can also indirectly provoke a local collective actions in favor of a stricter policy for the preservation of woodlands. Local inquiries paid also attention to the topic of farmland consumption and preservation of natural spaces.

The conflicts we have examined and their evolution pertain to the manifestations of oppositions to land modifying projects as well as to the effective emergence of constraints, pollutions and nuisances related to the changes that have occurred in the original space. Thus, the emergence of conflict is not necessarily related to the occurrence of a material event, but can also correspond to an expectation by certain categories of actors opposed to the project that that event will occur.

The Participants to Conflicts: Different Actors and Combination of Actors

The individuals or organizations involved in land use conflicts can be divided into two large categories:

- Users of land and resources for productive purposes (whether or not they are the owners of the land in question and of their work tool): artisans and industrial entrepreneurs, farmers, herders and forest entrepreneurs, providers of recreational services involving the use of land;

- Users of land and resources for non-productive purposes (present on the land permanently - residents, hunters, sportspeople, hikers – or only present occasionally – tourists, secondary residents).

These categories of users find themselves involved in the tensions and conflicts, either individually or as parts of networks or groups of actors. Conflicts can involve users opposing one another (whether or not they pursue identical production goals), reveal oppositions between different categories of users. However, we consider that many actors can combine productive and non-productive functions, thus going beyond the simplistic dichotomy between the ones and the others. Indeed, they reveal contemporary social complexity and the various roles one person can play.

It is for this reason that we have chosen to base our work methodology on actors rather than on the uses they make of the areas considered. We also use the term "actor" so as to avoid referring to large categories of land users (residents, farmers, environmentalists, industries....) which are abstract and often account for only part of the reality and complexity of the actors themselves and of their relationships with others. Following the example of Janelle (1977) and of Ley and Mercer (1980) we also use the term "conflict participant" - which is the base economic and social unit of conflictuality - or conflict stakeholders.

The Motives of Conflict

Conflicts arise from changes or projects of change, perceived by some actors as contrary to their interests or their wishes. The material expression of changes at the origin of conflicts pertains to several categories:

- Construction, deterioration or destruction of property, a landscape or infrastructure;
- Creation of a new production facility or expansion of an activity;
- Emission of external negative effects (diffusion pollution, odors, water drainage);
- Development of a property or piece of land;

- Access issues (restriction/exclusion, or opening/easement).

Nevertheless, the properties or facilities do not necessarily have to actually exist for a conflict to emerge. Conflicts can result from projects of construction, of implementation or extension of an activity, of development or of access modification, or from the emission of negative effects. In this case, conflicts are considered "anticipative" or "preventive". Indeed, disputes may arise between local residents or associations (acting individually or together), on the one hand, and the administration and applicants for permits to develop building projects or to operate activities, on the other. These disputes concern application for permits, above all, and come into play *a priori*(preventive conflicts), i.e., before the environmental risk-bearing activity or development project is actually launched. In this case, the dispute arises at the outset of the process. However, disputes may also arise when industrial operators or developers challenge administrative decisions or when local residents protest against the consequences of pollutions, nuisances and environmental impacts and the dispute occurs then *a posteriori* (curative conflicts).

From Tensions to Conflicts

The distinction between tensions and conflicts is tricky to analyze. Indeed the emergence of a conflict follows an explicit engagement of the actors; an engagement which takes the form of acting out episode: threats, assault, actions at law, technical acts, and signs (forbidding access....). We shall call conflict any tension that turns into a declared confrontation, via the engagement of one or several parties.

We then consider that tension between various parties designates an opposition without the engagement of the protagonists, whereas a conflict emerges with the engagement of one of the parties. This engagement is defined by the implementation of a credible threat, which may take different forms:

- Legal actions
- Bringing the matter to the attention of the public authorities or of the civil service representatives
- Mediatisation (bringing the matter to the attention of the media, press, radio, television...)

- Assault or verbal confrontation
- Putting up signs (signs forbidding access, fences and gates...).

More or Less Apparent, Individual or Collective Manifestations of Conflictuality

Sporadic or recurrent, the tensions and conflicts related to the different land uses can manifest themselves in various ways:

- At inter-individual level: bad relations between neighbors, assault, recourse to third parties, retortion, reprisal;
- At a more general level: carried or handled by individuals (elected officials for example);
- Finally, at collective level, carried or handled by groups, particularly associations representing actors using land for non-productive purposes (these groups distinguish themselves from enterprises or large scale farmers or forest entrepreneurs in that they are characterized by a non-hierarchical internal organization and non-productive purposes), administrations, local or territorial authorities.

The strategies of the groups and individuals are closely related to the conflictual events. Highlighting them helps to explain the objectives and positioning of the actors at the onset of the conflict and in the ways to manage it. The way tension or a conflict is managed often depends on the goal pursued.

Tensions and conflicts have histories. Some conflicts pass away quickly, whereas others can last for long periods, with phases of more or less intense confrontation and stages of more or less latent antagonism. A phase of tension can last for a long time without turning into conflict, if the actors do not engage. During the period of tension preventive actions can be undertaken in order to prevent a conflict from happening. Nevertheless, it is important to note that it is perfectly possible for a conflict to arise without their having been prior tensions.

Modes of Prevention and Management of Conflicts

Although we make no hypothesis about the necessity to solve conflicts we regularly meet actors who seek to promote or implement modes of conflict resolution. The modes of conflict prevention and management can:

- consist of preventive actions aimed at promoting appeasement and at avoiding the occurrence of a conflict. These actions can be taken so as to promote inter-individual negotiation; they can consist in involving a third party representing land users, or in encouraging the actors to adopt a non-judicial route such as institutionalized mediation for example;
- merely consist of an arrangement between the actors involved;
- rest on a regulatory or legal techniques.

The Solutions Considered

Without going into an in depth description, we identify below the main types of solutions implemented:

- Technical acts;
- Compensations (financial, natural or technical);
- Land use planning;
- Eliminating the activity from the area in question or moving it somewhere else;
- Adjudication;
- Mediation by an insurance company.

Methods of Conflict Identification

The identification and analysis of conflicts rest essentially, in our method, on following sources of data collection:

- The daily regional press (DRP);
- Surveys conducted by experts;
- Data from administrative litigation courts;

- Data through other sources are often added in studies on developing countries, where some of the previous relevant sources are missing;
- Internet (personal blogs and web pages)
- Geographic Information System (GIS).

Every source of data collection has its own limits. Only a small part of the conflicts (e.g. regarding the engagement paradigm previously identified) is treated by tribunals and a lot of opposition does not go on Courts. As far as the political traditions and the laws are different in each country, tribunals can play different and more or less important roles. The judgment can also be subjected to various interpretation or contestation, even regarding the administrative component. In the same spirit, there are severe limitations to the use of DRP. First, it is obvious that media behave differently with regards to the countries but also the regions or the areas within a given country. For example, they do not behave the same way in the north or the south part of France because of local traditions of public expression. It is also true that some media lie or forbid crucial events, or even at least euphemize about certain events, in order to hide a part of them or to reduce their importance. The same applies to experts opinions: they can lie, cheat, euphemize, or sincerely forget some crucial events. For all these reasons, we have chosen to base our identification and our analysis of conflicts on the triangulation of three main sources of data. Each one has its own limits, but altogether they provide a credible image of a conflict, at least regarding to the above definition.

A conflict is a social construction, which is identified by the observer and depends on the definition of conflict events and the types of observations. In our method, the identification of the conflict is based on a triangulation of the main sources of information (DRP, Surveys, litigation data or other data). It is by cross checking and comparing the various sources of information that an evaluation of the state of conflictuality in a given area is made. This job is performed by the experts of the team, on the basis of the main data. The status of "conflict" bears on at least two identifications in the main corpuses of data, for example in the daily press and regarding verbatim by local experts. The types of conflicts identified and the comparisons between different data bases aim at identifying several discrepancies between local situations; for example, DRP helps in revealing early oppositions to local projects whereas administrative litigations are related to well informed situations and revelation of infrastructure projects by mainly public authorities.

The identification of the conflicts, is however, preceded by the identification and diagnosis of the study area. Indeed, we focus on precisely delimited geographical areas and we collect and analyze the data pertaining to conflicts occurring within those zones. This enables us to describe with precision the conflicts and their evolution, and limits the number of possible occurrences.

Identification and Diagnosis of the Study Area

The area examined is always situated within an institutionally determined zone. Its geographical delimitation rests on that of the local public institutions, such as: communities of municipalities, conurbation comities, counties, regional nature parks, sub-regional constituencies.... The study area can comprise one or several of these administrative zones – several constituencies for instance. We shall only examine the events that occur within the chosen area. An exception is made for litigation studies. Because the amount of data on the rulings of secondary tribunals is usually too small and therefore not representative, we examine the district or region in which the study area is situated.

Once the study area is defined we perform an area diagnosis which must enable us to identify its main social and economic characteristics and its salient features and actors involved within it.

The basic diagnosis report, of approximately 10 pages in order to make comparisons, must comprise:

- A general presentation (location, the geographic morphology, history, social and economic dimension…);
- A presentation of the activities that depend on the resources available on the territory;
- The main territorial governance structures, particularly the institutions.

The Daily Regional Press (DRP)

DRP reaches in the hands of about 400 million readers around the world every day. It is the second most popular medium after electronic media, and therefore an interesting tool of observing voices of regional population. Furthermore, it has the twofold specificity of being the main medium of local news delivery and, for each daily regional paper, to often enjoy a quasi-monopoly in their circulation area. The DRP articles are an easily accessible source of information on pre, during, and post-conflicts, and complements efficiently the data found in other sources, through surveys in particular. The DRP provides relatively detailed information about local affairs, which is not the case of national newspapers (Rucht and Neidhardt, 1999; Mc-Carthy et al. 1996).

The work consists in gathering the information provided by a given daily newspaper, by examining all the issues published over a given period, of at least one year. The newspapers can be consulted online or in paper from, depending on availability. In the case of online consultation, the inventory can be conducted via the dissemination server providing access to the digitized editorial articles published by the main daily national and regional newspapers. The automatic keyword based search method is only used when information is sought about a particular topic, and not when a general inventory of all land use conflicts is needed. All articles are then displayed one by one, before being selected or not as part of our corpus. The articles quoted in the following sections were selected using criteria that have enabled us to differentiate the situations of tension from situations of conflicts (See Section Definition of the scope of investigations).

When the information contained in one article enables the researcher to identify the action, engagement and connection of an actor, or when the article provides complementary information about a previously described conflictual situation, it is indexed in the corpus using different variables:

- Its title
- Its date of publication
- The issue of the newspaper
- The section and page in which it appeared

- A very short summary of the facts described

Whenever possible a copy of the article is made (otherwise the summary replaces the copy). The list and the copies of the articles enable us, in laboratory, to group together the articles relating to the same conflict.

This process is not aimed at describing exhaustively all the conflictual situations. Rather, we deal here with a specific type of event, which is the one written about and reported to the public through the press. This source has important biases which rules out using it by itself. The press omits some events; it can have a tendency to euphemize, dissimulate, it can be partisan or controlled by certain interests. However, using the press has for several years been recognized as a means – in the perspective of an exhaustive quantitative analysis of conflict – of obtaining the "most complete account of events for the widest sample of geographical or temporal units", as Olzak (1992).

Surveys Conducted by Experts

Interviews with experts provide information about the level of conflictuality. We aim to cross check this information with other sources, or to highlight conflictual events that might not have been brought to the attention of the courts or of the DRP. The interviews are aimed at identifying, in each zone, the dynamics of evolution of the rural and peri-urban areas concerned, at determining the types of conflicts and tensions relative to competing uses of rural spaces and at finding out the solutions implemented in terms of territorial governance.

We take great care of the fact that the experts interviewed have no private interest on land or other local resources involved in the conflicts, so that there are not personally affected by changes in land use for example. For the case studies quoted in next section the interviews were conducted with experts contacted beforehand by telephone, using a list of 40 to 50 people per study area. We choose to interview experts from various professional and associative fields so that a wide variety of opinions were represented. Each interview session (face to face) lasts between two and three hours. Each expert can be interviewed once, or several times when necessary (depending on the sufficiently required information). The respondents were interviewed with an open questionnaire so as to obtain as much information as possible, concerning conflicts and their evolution.

The questions may not directly be related to the conflicts. Indeed it has been observed that asking direct questions about conflicts generally leads the interviewees to refuse to answer. The interviewers present themselves as conducting surveys on local situations of governance, of actions and interactions between actors, sometimes with specializations depending on the interviews and the institutions they represent. The questions are always indirect: the interviewer must be trained to identify the conflictual elements of a situation. They must ensure that the questions in the grid presented below are all answered. The analysis of the conflicts is conducted later, in committee meetings.

This work enables us to gain a more thorough understanding of conflictual processes, to describe them and to identify their components:

- The materiality of the conflicts;
- The actors of the conflicts;
- The motives of the conflicts and how they emerge, which contribute to generating them;
- The manifestations of the conflicts, which imply various levels of symbolic or effective violence, the engagement of the actors ranging from petitioning to engaging in legal proceedings, or assault.

The 10-page report of the conflicts analyzed on the basis of the interviews with the experts lists the main conflicts that have occurred in the area considered, according to the interviews conducted. Quantification is impossible but the report provides insight into the perceptions of some of the key local actors on the conflictual process. A short one-page synthesis is provided with the report summarizing the main information obtained.

The interviews with experts cannot be our only source of information about conflicts for they present strong biases due to the method of analysis: the actors may have forgotten some elements; they may amplify or minimize, omit or lie about certain points. It is therefore absolutely necessary to complement them with other sources of data. Nevertheless, the interviews provide information that other sources cannot provide, and help us to gain insight into the dynamics of local alliances and oppositions, by enabling us to interact with the actors of the conflicts or with the observers of conflictual situations and of their development in the long term. We agree, here, with the conclusions of other studies about conflicts drawn on the basis of interviews with local actors, and more particularly with the conclusions of a study conducted for the World Bank by Deininger and Castagnini (2006) about the conflicts related to land property in Uganda.

Thus, we conduct semi-directive interviews.

Interview and Interview Analysis Guide

The interview guide indicates the different items of information that must be collected in view of the analysis and exploitation. It does not provide with a clear description of the questions that must be asked during the interview. The interviewer must ensure that all items are discussed before ending the interview and must write a report. The questions to be asked and answered are the following; the order of the items is only indicative.

- Location (precise or extended)
- Spatial support of the conflict (punctiform or linear)
- Activities and their restrictions of use (productive, residential, recreational, "nature", exclusion, network infrastructures, public facilities)
- • Number of actors (or groups) involved
- Actors (or groups) (by analyzing their degree of organization); distinguishing the actors who have an institutionally recognized mediation functional
- Origin/trigger of the conflict
- The causes mentioned
- Relation to time (duration, frequency)
- Relation to space (evolution of the space of concern during the conflict, by distinguishing if necessary, the places of litigation, the places where the conflicts take place and the places that are evoked by the actors)
- Forms of expression: (a) Verbal expression, individual expression (threatening mail, altercation, intentional damage of property, confiscation) collection expression (petitions, pamphlets,

collective occupancy, marches, etc.). b) Departure (eviction, non-participation, passive refusal to obey a legal order or a summons, lasting refusal to participate. c) legal action (civil or administrative jurisdiction)

- Demands/claims declared or not by the protagonists (elimination or minimization of the nuisance – with or without a proposal of technical solution), demand for material or symbolic compensation, interruption of construction, etc.)
- Development (possible solution, the conflict continues, agreement, tribunals....)
- Public facilities possibly cited during the conflict (as a cause, as an element of context, as a solution)
- Types of arguments put forward:
  1. effects on the personal living environment, versus activities, versus natural environment, collective, on the health of the individual or of the group, on the personal versus collective costs
  2. on the principle: transgression of a locally accepted rule of a group, of an official law
- Evaluation of the behavior of the actors who are statutorily assigned to intervene as mediator, as guarantors of the rules

We refer here to the list of local experts. They are obviously not all present on all the territories, but the goal is to provide information about each main category of the preceding Box and to achieve a balanced representation

The Experts to be contacted on Each Site

Local Public Institutions

- *Local elected officials*: mayors of municipalities and other elected representatives (General councilors…)
- Directors/managers of inter-municipal services/facilities
- Economic director of inter-municipal planning
- Chairman of an inter-municipal body or of specialized committees (environment, tourism, agriculture, commerce…)

Institutions for the protection of the environment or organizations of nature users

- Regional Directorate for the Environment (environmental protection agency)
- Local associations for the protection of nature or for the environment (water, hikers or other outdoor sportspeople)
- Federations of hunters and fishermen and their local associations
- Environment and Energy Management Agencies

Forestry and agriculture related organizations (Specific attention is paid to this area of activities, and particularly on agriculture)

- National/regional forests offices
- Federation of Forestry Owners
- Local Land development and management bodies
- Chambers of agriculture
- Planning departments and agricultural economics services
- Federation of Farmers Unions

Regional directorates (planning, economic services, agriculture and forestry)

- Farmers (in person)...
- *Socio-professional representation*

Chamber of trade

- Chamber of commerce and industry
- Entrepreneurs clubs
- *Infrastructure planners*
- Water agency

Agency for electrification and water management

- Agency for highway/motorway construction
- Land Development and Rural Settlement Associations
- Land acquisition and relocation committees
- Department of railways
- Local construction agencies
- *Other State services*
- Sub-prefectures (general secretaries or attached functionaries)
- Economic services
- District Court clerk

- *Other sources of information*

 Local press journalists
- Notaries
- Judges...
- Police department

Analyses of Litigation Rulings

Administrative litigations should be collected from respective courts for statistical analyses. Therefore, statistical analysis of judiciary sources aims at examining the way in which legal rules are mobilized in land use conflicts, using a study of litigations between the parties involved in conflicts. This analysis targets a specific category of conflicts: those that have taken a specific path and have led to legal proceedings.

The inquiry on conflicts, through the quantitative study of litigation, is an empirical methodology that has its own limits and opportunities. As for the limits, the research in legal sociology usually concludes, regarding empirical data, that the resort to courts is always intrinsically marginal, may the phenomenon be observed in developed or developing countries. The successive social acts that consist to get informed on its rights, to choose a legal advice to solve the dispute and to go to the courts constitute different stages that select the people involved in conflicts. But at the same time, even though the actors reluctant to settlement and going to courts are a small part of the whole set, this tiny fringe is likely to be a mass phenomenon concerning a huge quantity of cases that can be studied statistically. Furthermore, the study of this small fringe of litigants can inform us about non judicial settlements: for example when cases are filed but dropped before they are adjudicated, it may be a sign that the resort to courts was just a mean to open a negotiation.

In some developing countries, this selective dimension of litigation is surely more acute. Cultural and financial barriers for the access to courts, the likelihood of corruption among the judges, the weak probability of implementation of court decisions, given the failure of public administration, are disincentives for a judicial strategy. Nevertheless, if the social phenomenon of litigation is quantitatively less important in those countries, our researches (conflict over water dam in Pakistan) and other empirical inquiries (for example, the world survey of Antonio Azuela for the Lincoln Institute about the conflicts over land evictions) (Azuela and Herrera-Martin, 2010) show that the dynamic of the process can be favorable to a development of judicial strategies. In that case, the research will focus more on evolutions that indicate an increase of legal consciousness than on the intrinsically low level of judicial disputes on the short term.

In France, disputes concerning activities which bear environmental risk or are likely to impact farmland or preserved natural spaces are adjudicated by Administrative Tribunals. These frequently involve claims initiated by associations and local residents against projects which have been granted permits by administrative authorities but whose impact is alleged to be harmful to the health and well-being of the local environment and community.

The various forms of actions at law have also been the objects of fruitful investigations in matters of infrastructure development or urban planning. The study of the practices in penal law in this field reveals the maneuvers possibilities that exist at all stages of the treatment of the offence, and the possible solutions: by the administrative authorities that detects the offence, by the prosecutor who decide to not prosecute, or by the judge who decides to pass an alternative sentence. They also highlight the diversity of the origins of the offence reports submitted to the legal system, depending on the priorities determined locally by the administrations concerned – targeted prosecution campaigns over a given period so as to solve a local problem – in a field where offences are much more often reported by booking officers than by the victims of the offences (Struillou, 2004).

With the exception of rare studies targeting specific associative actors, (Leost, 1998), the statistical studies undertaken on the basis of data from administrative jurisdictions are almost non-existent, in the field of land-use litigation. These sources are only used in the framework of general statistical studies on the activities of administrative tribunals, which, however, provide a useful basis for conducting more targeted surveys (Barré et al. 2006).

Therefore, the analyses of conflicts based on the observation of legal and administrative litigations are conducted at small administrative unit (*department, district, division etc.*) level. This choice is justified by two arguments: on the one hand, that particular administrative unit is the territorial unit of reference for many actors, whether they be public actors in charge of land use regulation (prefect, non-central government services), or para-governmental and private actors: Associations for the protection of the environment generally act at these administrative levels; the same applies to chambers of agriculture or the associations of hunters and fishermen; on the other hand, court rulings generally state precisely where the problem is situated at municipal level (private litigations, applications for annulment of municipal decrees or deliberations) or at departmental level (applications for the annulment of prefectoral decrees).

Land use and environmental conflicts consequently reflect the way this administrative activity operates. Knowledge of the inner workings of this system on a normative and empirical level is imperative in order to understand the logic behind the litigation. In the case of administrative justice, this prerequisite seems easier to meet than in other fields, like criminal sociology, or the sociology of civil disputes, to the extent that the dispute can be traced to administrative decisions and activity for which there are more likely to be written entries. This may not be the case for private law documents or contracts, so it is easier to obtain data on the frequency of claims in the case of administrative justice. It is therefore with regard to the intensity of conflictual activity must be assessed.

In developed countries, especially in France that the corpus of court rulings is built using the*Lamyline* textual legal database which comprises the complete texts of the rulings made by the Court of appeals and Supreme Courts. More precisely, this database comprises all the rulings made by:

- The State Council (highest administrative jurisdiction in France) since October 1st 1964,
- The Administrative Appeal courts since January 1st 1989,
- The Court of cassation (highest civil jurisdiction in France) since October 1st 1959 (excluding the rulings by the Criminal Chamber which are included from January 1st 1970 onwards),
- The Court of Appeal since January 1st 1982.

The software has a search engine that enables us to use the Boolean operators and those of several French case law data bases. The search for rulings by the four levels of jurisdiction defined above is conducted from the following case law data bases:

- State Council
- Administrative Appeal Courts
- the Court of cassation
- Appeal Court

Contrary, in the developing countries it is often difficult to collect the statistics from administrative courts, because of the absence of case law databases. Thus cases of the courts can be collected by searching manually from the yearbook records in their libraries, which is of course a time taking process. While in some developing countries uncooperative behavior of the administrative courts is discouraging the researchers[a]. Although the case law hierarchy is not the same in the entire developing world, it follows more or less the following structure:

[a]Forexample,in Pakistantribuna lsrefused to providedata on land use conflictrelated cases because of involvementofbureaucrats,feuda landpoliticians.

- Supreme court
- High court
- Session court

Obviously, only part of the land use conflicts has been analyzed: those that have been dealt with by Courts of law, which limits the number of conflicts selected and examined. Indeed, the "passage through court" is the result of a selection using various filters: an individual's refusal to negotiate, an administrative authority's refuse to regularize… or on the contrary the intention to use the tribunal as a lever for starting stalled negotiations or for conducting them with a more favorable balance of power. These elements contribute to giving the conflicts that have been dealt with in court a profile whose specificity must be discussed in the interpretation of the surveys' results, without however considering that they represent all the conflicts that occur in the various territories.

Data from Other Sources

The identification and listing of the conflicts are performed on the basis of the data from the three sources described above and on their comparison.

Nevertheless, other secondary methods of analysis can be added to this initial list, especially in developing countries or less developed countries, where some of the previous sources of data are missing. This can be particularly true for litigation data. These additional methods are aimed at conducting a more precise analysis or at specifying such or such dimensions of conflicts.

We shall, here, content ourselves with mentioning them. They are respectively:

- Sectoral analyses: for example, analyses targeting the farming sector or land property issues;
- In depth analyses carried out by researchers/experts in specific disciplines: for example, interviews of actors by sociologists, or monitoring of meetings or postal questionnaires administered in the context of research in social psychology;
- Study of particular situations: for example, study concerning a river, or farming facilities or land takes for road and water reservoir construction.
- In depth analyses of digital aerial photographs and remote sensing images through Geographic Informative System (GIS) of the territory; for example if the changes over a piece of land are being monitored in case of superposition of uses and its impacts;
- Analyzing personal blogs and web pages through internet, which is growing source of information on opinions and expertises in the developing countries.

Inquiries implemented in Greece combined different sources, from local newspapers to online websites supported by environmentalists associations ("observatory of natural spaces"). The latest source has the advantage to cover a large period (2001–2011), but is characterized by a selective approach, since data are furnished by local activists. Other sources complete the research methodology as follows: interviews with local experts (journalists, lawyers…) and key stakeholders like planning agencies, political leaders and elected officials, additional events collected through other web platforms (blog sites of environmental activists). An important source of documentations was also constituted by political materials like programs of local parties for municipal elections.

For analytical methods of the case study from Pakistan (presented in next section), data have been collected from various secondary sources to crosscheck the results. For example, the information has been collected from literature published by public and private organizations, on the case study perspective. The data have also been collected through aerial photographs and remote sending images, through satellites which were treated under GIS in order to examine land cover changes and impacts of the superposition of land uses due to decision for infrastructural projects by the public authorities over existed economic activities in the region. This technique of research became reliable after development in technical software. Moreover, data also collected after in-depth analyses from personal blogs and different web pages, these sources of information are very promising for future researchers of land use conflicts due to huge development in internet in the developing countries.

The Conflict © Database

The scope of our method is rather large, and we intend to provide it as an opportunity to studying land and resources conflicts in developed as well as developing countries. However, we have built a data base related to our own investigations, especially in France (13 case studies in different areas) and Pakistan. This data base can be expanded or reproduced for other types of investigations, in other countries or regional areas, given its wide scope and the general character of the contained items. We will present it in in broader terms in the following paragraphs.

In the context of the research conducted on conflict, using the relational database responds to a twofold analytical need. The aim, first of all, is to create data that can be exploited quantitatively, using the documents and survey results: the coding operations performed on the basis of the source documents are aimed at quantifying the phenomena of conflictuality, which will be analyzed and possibly examined from the perspective of the profiles of the areas concerned. But this database must also enable us to conduct a comparative evaluation: comparison between the sources used (court cases, newspaper articles, questionnaires), and comparison between the different areas surveyed.

The comparison between the different sources rests on the database' structure, which is designed in such a way as to be able to compare them. Besides the variables relative to specific observation contexts (type of rules used in the cases, number of press articles discussing a particular subject) some common variables have also been defined. The researchers conducting the coding operations must translate the specificity of their material in these cross sectional categories – described in detail below – relative to the types of conflicts observed, to the actors involved, or to the uses and arguments discussed.

Database Structure and Categories of Conflicts

The overall structure of the database can be schematized in the Figure 1 below. The data base incorporates three main data tables, that is, in order of inclusion:

- A table containing the variables relative to the geographic locations of the conflicts (in relation to a municipality, a community of municipalities, a *département* or a district);
- A table indicating the variables describing the conflicts per se, that is, the cross sectional categories – which are identical whatever the source of the data, and the categories relative to a context of observation (The legal categories defining, for example, the nature of a request made to a jurisdiction);
- And finally a table providing information about the profile of the actors involved.

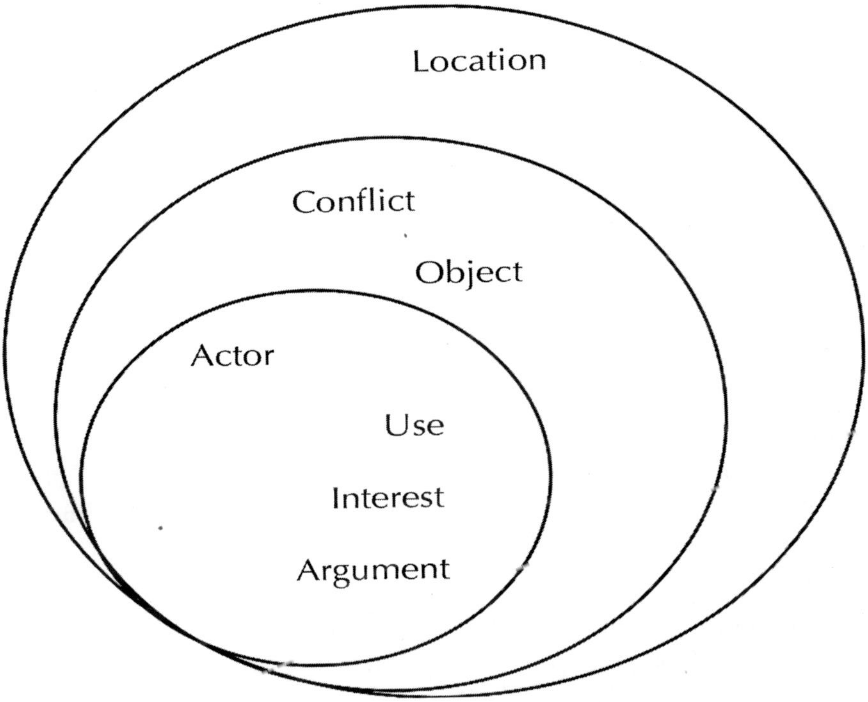

Figure 1: Simplified graph of the database' structure. The circles represent the relationships of inclusion between the tables. We have noted in italics the tables that are dependent on the "conflict" and "actor" tables.

Four other dependent tables must be mentioned: the description of the conflicts' "object", as well as information about the actors concerned: what the actor involved in the conflict uses the land for, the arguments the actor uses, and the interests that motivate the latter's engagement in the conflict.

The difficulty of studying conflict in one territory using different types of sources has to do with the identification of comparable analytical categories, whatever the observation point selected. This difficulty is crucial in the analysis of the requests sent to tribunals, in so far as those requests are expressed in a constrained language, that of the legal categories. Moreover, the enumeration of the criteria defining the contours of a conflict as a homogenous entity is a methodological demand which becomes particularly acute when exploiting press articles. Thus, while "formatting" the actions at law as clearly identifiable cases naturally leads us to choose "case" as the unit of measurement of the "conflictual" event, it is the "article" that is used as the measurement unit for the analysis of the press, when reconstituting the event after it has occurred.

Furthermore, a common definition of categories that cross the analysis of actions at law and that of the press, has to be found in order to enable the researchers to use a common language of description of conflicts. This does not exclude the necessity to define, for each type of observation material, specific variables relative to the source used. Thus, the nomenclature of requests, used by administrative court clerks for example, has been used for internally describing the different types of actions at law. Similarly, some variables are only meant to be used in the framework of an analysis of the press (number of articles devoted to a particular conflict). We present here the main components of the database, which is described comprehensively in several internal documents (for example, Galman and the participants of Conflict Program 2007).

The trickiest variables to define are, naturally, those that pertain to the different objects of conflict (see below). The detailed examination of the categories defined reveals the combination of several modalities of definition; complexity inherent to the expression of the forms of conflictuality. Indeed, the "objects of conflict" refer, depending to the cases, to:

- more or less clearly territorialized economic activities, whether they specifically involve using natural resources (agriculture, extraction of underground resources), location of production activities (industrial production, production of energy, waste treatment), or activities related to the presence of amenities (tourism);
- types of legal authorizations granted by administrative authorities, in as much as they correspond to land uses as defined by law;
- forms of social relationships marked by geographical specificities: Neighborhood relationships for instance.

The Objects of Conflicts: the Categories Used

- *Accessibility and easement*
- Right of access and passage

- Occupancy/parking
- *Facilities classified for the protection of the environment*
- Quarries, borrow pits
- Salvage/recycling, storing, waste treatment
- Salvage/recycling storage of materials (cars, tires…)
- Production, storage of chemical substances
- Other regulated industries
- *Site rehabilitation*
- Mining site
- Industrial production site
- Storage site
- *Service activities*
- Tourism, leisure
- Transport, Fuel distribution
- Trade, retail, advertising
- Road, rail, sea, air transport
- *Agriculture, forestry, fishing*
- Farming
- Forestry
- Fishing
- *Public utility infrastructure*
- Airport, railway, road, harbor infrastructure
- Construction of water reservoir
- Production – energy transport
- Telecommunication infrastructure
- Public facility
- *Management and Conservation of Natural Areas*
- Hunting/fishing
- Water quality management
- Soil quality management
- Air quality management
- Landscape, wetlands
- Conservation/ management of the Fauna/flora/biodiversity

- *Urban planning/development operation or document*
- Camping sites
- Urban development document
- Right of pre-emption
- Public domain occupancy
- Land restructuring
- Risk management areas
- Natural areas
- Management of natural areas
- Construction/extension of agricultural buildings
- Construction/extension of residential building
- Construction/extension of commercial building
- Construction/extension of thoroughfares
- *Neighborhood*
- Neighborhood nuisances
- Declaration of co-ownership
- Theft, damage, attack

Profiling the Actors Engaged in Conflicts: Categories of Actors, Land Uses, Arguments

The analysis of the actors engaged in conflicts is summarized with great care in the grid of the variables used. Once the credible engagement of an actor has been identified, one can, indeed, distinguish the actors that are at the origin of this action (the protesting actors) from those which this action targets (the disputed actors).

As stated in point 1.2., the typology of the groups of actors that we have used for the database is founded on a distinction between users of land and resources for production purposes, whether or not they are the owners of the land (farmers, forest entrepreneurs, artisans, industrial entrepreneurs, providers of recreational services...) and users of land and resources for non-productive purposes, whether they use the land in question occasionally or continuously (residents, hunters, fishermen, sportspeople, hikers, tourists, secondary residents...) (See below). If a local authority body, or a State department is, for example, user of resources (in the case of well boring for water for municipal consumption, or of public road maintenance by the public services), we shall consider the actor concerned as a "representative", considering that during a conflictual confrontation, s/he will refer to the authority which his role as representative confers upon him/her, and as defender of collective interests.

The Categories of Actors Involved in the Conflicts Studied

- *Farming or equivalent occupation*
- Farmers
- Irrigation operators
- Forest entrepreneurs
- Fish farmers
- Herders
- *Industrial actors*
- Artisans
- Industrial entrepreneur (mining activities)
- Construction and civil engineering firms
- Industrial entrepreneurs in the manufacturing industry
- *Actors of the market service sector*
- Transport enterprises
- Energy suppliers
- Providers of services related to waste management
- Providers of services related to water management
- Real estate developers, planners
- Actors of the hotel industry
- Other market services
- *Associations*
- Ass. For the protection of the environment (national or worldwide)
- Ass for the protection of the local environment
- Hunters and Fishermen's Association
- Political party

- Other associations
- *Local public authorities*
- Region
- Department
- Municipality
- Public Intermunicipal Cooperation Institute
- Regional Natural Parks
- Local Administrative body (water agency, Committee for the protection of natural sites and monuments)
- *National Public authority*
- Minister
- Prefect
- De-concentrated State services
- Judicial authority
- *Elected representatives*
- Municipal, departmental, regional representatives
- *Professional organization*
- Trade unions
- Consular chambers
- *Individuals*
- Permanent residents
- Temporary residents
- Owners syndicates
- Tourists, non-sedentary population

Once the actors of the conflicts are identified, we describe the controversial or conflict generating land uses that is the land uses the spatial consequences of which the objecting parties consider as sources of constraints. The distinction between actors and uses requires the taking into account, through standard observation, of the fact that one user can use land for several purposes: A farmer can also be a herder, a hunter, or a defender of nature; an industrial entrepreneur may also be a hiker; a resident may in the framework of his/her work conduct polluting activities...

Different configurations of opposition emerge, including between actors making the same use of land. Particular attention is placed on the local public mechanisms that are liable to exacerbate or crystallize certain tensions into conflicts.

The Categories of Land Uses Identified for the Analysis of Conflicts

- *Creation of infrastructure*
- Production and transport of energy
- Road, railway, airport infrastructure
- Waterway transport infrastructure
- Waste management
- Water supply
- Public buildings
- Telecommunication
- Advertising infrastructures
- Recreational/tourism infrastructures
- *Production of services and exploitation of infrastructures*
- Tourism, hotel and catering industry
- Telecommunication
- Air, road, railway, river, sea transport
- Production and transport of energy
- Trade, advertising
- Sanitation
- Domestic waste transportation and management
- Water catchment, treatment, storage and supply
- Management of hazardous waste and materials
- Utilization of buildings housing public services (functioning)
- Exploitation of recreational or tourism infrastructure
- *Agricultural, fishing and forest production*
- Agriculture
- Aquaculture

- Breeding
- Professional fishing
- Forestry
- *Industrial production*
- Mining
- Manufacturing
- Site rehabilitation
- *Recreational and touristic use*
- Hunting, fishing
- Hiking, motor sports, non-motor sports
- Tourism's events (cultural, musical, etc.)
- *Residential use*
- Construction/extension of social housing
- Development of poor housing
- Residential use
- *Conservation and management of resources*
- Fauna, flora, biodiversity
- Surface and underground water
- Site, landscape
- Ground
- Heritage
- Risks

Absence of a Characterized Land Use

Just like the distinction between actors and users, the identification of specific registers of argumentation is based on the hypothesis that some categories of actors can, depending on the conflictual situation, use specific arguments. The capacity for technical and legal argumentation can for example depend on the degree of collective mobilization which a given conflict gives rise to.

An important cornerstone for the analysis of conflicts consists, indeed, in highlighting the registers of argumentation used by the actors. Argumentative openness implies that the (individual or institutional) actors diversity their registers both from the point of view of values and that of rules, so as to lend as much efficiency as possible to their protest movement. It can perfectly well result in contradictions when the same actor puts forward both the individual's interest and the general interest. But beyond this classic opposition, these argumentative strategies appear to be more complex when using the "general interest" argument requires the mobilization of distinct norms, which are often revealing of contradictions in the law itself (Lascoumes, 1995): thus, in the context of a protest against an urbanization project, the principle of controlled development of land and the argument in favor of housing development are used in turns; both categories of general interest are recognized by legislation.

Registers of Argumentation: Categories of Analysis

- *Scientific and technical argumentation*
- Protection of ecosystems
- Infrastructures and equipment
- Risk evaluation and management
- Socio-economic argumentation
- Economic interest
- Sustainable development
- General local, regional or national interest
- *Reference to private land rights*
- Access
- Ownership
- *Responsibility*
- Private responsibility
- Public responsibility
- *Quality of life*
- Perception of risks
- Living environment
- Insecurity of individuals and property
- *Values*

- Modernity
- Tradition
- *Respect of the law and regulations*
- Risk prevention
- Other

Constitution and Exploitation of the Data Originating from the Daily Regional Press

After collecting the data from the DRP, the inscription is made in the data base using the following schemes. For each conflict identified, we include in the database a series of descriptive variables that groups together the following elements:

- A brief summary of how the conflict developed, as described in the relevant article or series of articles,
- The spatial environment of conflictual events, the facilities - being used or subject to usage restrictions – that are blamed as sources of constraints,
- The resources that have been modified, and the usage or activities affected by these modifications (reconstituted using the provided arguments),
- The opposing actors and their modes of engagement in the conflict,
- The geographic location of the goods/resources affected by the controversial facilities[b].

[b]One can associateto each conflictthe list of municipalities in which are located the sitesthatareatthecenteroftheconflict.

After analysis, we compile a 10-page long data sheet about the conflicts reported in the DRP. It lists the main conflicts that have occurred in the zone considered and classifies them according to the number of times they have been mentioned in the press. This helps us to obtain a picture of conflict, and of the media impact of the different causes of conflicts. This datasheet is accompanied by a one page synthesis summarizing the main information obtained.

Constitution and Exploitation of the Data Originating from Litigation Rulings

Once the database originating from litigation court ruling is compiled, it is analyzed statistically and lexically. The decisions are coded in such a way as to constitute a database integrated, first, into an Excel spreadsheet and then incorporated into a data processing software system (4D). The variables and modalities are defined using the above mentioned grid of analysis of conflicts also used for the interviews with experts and the DRP.

That corpus is exploited through the sequential and cross comparison of the significant variables, and through the analysis of the frequency of the legal references made. This analysis is completed by the textual statistical analyses performed with the software ALCESTE, so as to identify, through the language used in courts, the local specificities of the conflicts. These lexico-metric tools have only been used in the framework of the analysis of legal documents, which are well suited to this type of analysis due to the highly formalized structure of the argumentation. Without describing in detail the modalities implemented and the results obtained with this tool (Kirat and Torre 2004), it is useful to specify its scope as a component of a wider methodological approach to conflict analysis. The analysis of lexical statistics is justified by the fact that the language used in the court rulings is the product of the legislative or regulation texts used by the parties or the judge; this terminology is used to describe objects of protest or demand but also to express logics of action and goals (obtain the re-establishment of a right-of-way easement, the cancellation of a building permit in a zone of ecological interest, or the cancellation of a public inquiry conducted in the framework of a project of creation of a Classified facility for the protection of the environment). The lexical analysis is one of the elements of the study of the arguments mobilized by the actors. It enables us to show that one generic type of conflictual factors does not give rise to one unique and universal model of action in administrative courts. In other words, the actors seeking justice do not behave in a homogenous manner in all the *cases*studied.

The use of ALCESTE is not a technique for a-priori hypothesis testing, but for exploration and description. The program generates an empirically based classification of text units according to the pattern of co-occurrences of word tokens within these units. The lexical analysis software operates by dividing the corpus in elementary context, that is to say, a group of words identified by their length and punctuation. The similarity of these sentences is determined by reference to the similarity of the words used. A descending hierarchical classification distinguishes several classes of context units, and measures the closeness or distance between them. The classes identified constitute a basis for a principal component analysis (PCA), which is then used to realize graphical projection structured by the two most significant vectors. The researcher then begins to make interpretations based on the output from the program. It facilitates the discovering of the salient meaning structures and implements mechanisms for an independent analysis of the meaning of words, in order to get a statistical ranking of lexical statements in a given corpus.

The search for court rulings is performed by crossing the name of the departments selected with several keywords, defined in such a way as to cover as comprehensively as possible the scope of legal questions in which the land use conflicts can be formulated.

The Search Keywords Used for the Compilation of a Sample of Cases

- Easement
- Hunting or hunting right
- Building law
- Farm laws and hunting
- Farm laws and protection of nature
- Co-ownership or nuisance
- Directive 92/43/EEC
- Land application or breading
- Fauna and flora
- Classified facility
- Disturbance
- Nuisance and olfactory
- Nuisance and noise/sound
- National Natural Park or Regional Natural Park

- Passage and (hiking or Moto or quad)
- Rural land reparcelling
- Normal nuisance or neighbourhood nuisance
- Wetlands or swamp or bog

The judgments made in cases that occur locally at departmental level can be identified thanks to the effect of standardization of court decisions: the latter must include the address of the parties involved in the trial. Moreover, most decisions include the name of the study zone.

For example, the text of a judgment rendered by the French administrative litigation court is systematically made of four sections. The first states the identity of the petitioners, the nature and date of issue of the contested administrative act, and the administration that has issued the act, which is therefore sued in the litigation. The following section describes the judge's answer as to the means of form used by the different parties, that is to say the arguments relative to elements of the procedure (receivability of the request, the petitioners' right to act, etc.). The third section contains the judge's answer concerning the substance, that is to say the arguments used to oppose or to defend the administrative act itself. The fourth and final section evaluates the sanctions and compensations to impose depending on the ruling given.

At the end of this search, the corpus is reviewed so as to eliminate duplicates; indeed strong redundancy between the decisions found using different keywords can occur. The non-relevant decisions are subsequently eliminated, such as for example, the judgments concerning hunting accidents, or those relative to cases occurring in departments other than that where the parties reside...

Once the data is analyzed a 10-page datasheet on the conflicts identified through the exploitation of the litigation courts' data is compiled. It describes in detail, for each type of jurisdiction concerned (administrative and civil), the categories of requests that are the most frequently submitted to tribunals, and also indicates the main argumentation strategies observed according to the categories of actors (what corpus of rules is mobilized by what type of actor?). The question of the "outcome" of the conflicts, sometimes difficult to evaluate when analyzing the press, is here systematically interpreted, in so far as with the exception of abandonment or discontinuance of a procedure, a case judged on the basis of its substance gives rise to a decision that will oppose a "winning" party to a "losing" party. Thus analysis of the "rate of success" per category of conflict and per type of actors is therefore an important component of this synthesis. As a matter of fact, as in the case of the press and the interviews with experts, a one-page synthesis summarizes the main information obtained.

Added Local Socio-Economic Data

In addition to these elements we include in the database information relative to the area and municipality concerned and useful for understanding the local context in which conflicts emerge. These data are of two types:

- First of all, they are socio-economic variables describing the profile of a territory from the point of view of social dimensions (tax related data, proportion of social housing...), of environmental issues (percentage of farm and natural lands, of areas protected for their heritage value), of the demographic dynamics (migration, population pyramid), etc.
- A second group of data provides information about the different administrative decisions that are liable to give rise to opposition: Building permits granted by mayors (data on the authorizations granted by the Regional Directorates of Infrastructure), or permits issued by the prefectures in accordance with regulations *concerning classified facilities*.

Both types of data refer to two levels of explanation of conflictuality: The first, immediate, level of explanation helps us to evaluate a rate of conflictuality in relation to an activity of reference that generates controversy. Thus, intense conflictuality concerning urban development generally reflects a highly dynamic construction market. The succession of protests related to pollution issues is often the consequence of an area being highly exposed to nuisances because of the important number of classified facilities. However, the intensity of this conflictuality can be, in relative values, higher or inferior to that of the activity of reference. Using data on local social and economic background enables us to test hypotheses on the "long term" relationship between the dynamics of a territory and the modes of conflictuality, and to incorporate social and human dimensions. Thus, the level of conflictuality often proves higher in areas where the levels of income and education are high, which implies that the population mobilized is well informed and educated.

Illustration of Data Crossing Using the Conflict Method. Examples from Developed and Developing Countries

The use of our methodology of analysis, and of the data contained in the *Conflict* © database led to several works[c] (some of them, marked by asterisk * are presented in the bibliography), performed on a few case studies, and to an increasing number of publications, quoted above. Our aim, here, is not to enter into all the details of these studies, but to provide examples of studies based on different areas, in quasi opposite situations. Thus, we have chosen two examples, based on two case studies that have been studied extensively by our teams, in order to provide information on the way the method and the data base have been used but also about the options provided by them in front of different situations. The two case studies (the Greater Paris region and the Chotiari reservoir case in Pakistan) are rather different in many ways:

[c]The data contained in the data base are accessible to the interested readers if they want tocontacttheauthors.The detail edstructure of the databaseisalsoavailable(GalmanandtheparticipantsofConflictProgram2007)

- The Greater Paris region case study is located in a developed country, in a very densely populated area, in a peri-urban and very rich region, involving rural and urban inhabitants, conflicting for scarce resources;
- The Chotiari reservoir case study is located in developing country, in a sparse area, in a rural and very poor region, involving local users of land in a conflict against the loss of their land and natural resources.

Thus following subsections are devoted to the presentation of the dynamics of land use conflicts in these two areas, involving reflections on the use of our method in developed and developing countries, and about the possible adaptations to changing situations. We have also highlighted the different results provided by the use of various data, and their comparative utility.

Conflicts in Developed Countries. an example of Analysis of Conflictuality in the Greater Paris Region

The research conducted using our methodology in developed countries has led to studies of several sites located in rural and peri-urban areas in France[d]. They have resulted in several publications that jointly or alternatively use one or several of the aforementioned sources. They are of great importance with regards to the French case, because there is an ongoing debate on the role of land-use and proximity conflicts in rural and peri-urban areas. Namely: which types of conflicts occur in these zones? Are they related to public or private actors? Are they legitimate or are they part of the Nimby phenomenon? Can they be considered as signals of public decision failure? Do they prevent local development?

[d]TheSeineEstuary,theLoireEstuary,theRegionalNaturalParkofMonts d'Ardèche,thePaysVoironnaisinIsère,theCommunityofmunicipalitiesof MontrevelenAin,theCortenais,andBalagneareasinCorsica,thePuysmou ntainsinAuvergne,theRéunionislandarea,theGreaterParisRegion,theCh arentecatchmentarea,Montpellier'scoastline,andArcachonBay.

An illustration of how we have used the data collected using our method, performed by using the*Conflict* © database, is provided through the analysis of the levels of conflictuality in the Greater Paris Region, one of the areas surveyed (Darly 2009; Pham and Kirat, 2008; Pham et al. 2012), and, more particularly, through a comparison of two sources: data on the activities of the courts and the daily regional press. Paris is the national capital region of France, but also, and by far, the largest metropolitan area in France and can only be compared with two or three other metropolitan regions in Europe. Around the highly urbanised core composed of Paris and its suburbs, a peri-urban belt has received many residential and industrial activities that produce more or less urbanised rural landscapes. Within this peri-urban belt, the scarcity of well-located vacant spaces (well-connected to transport facilities and services centres) and the diversity of actors and interests that share the same rural environment raises many tensions and conflicts over farmland uses. But the peri-urban belt is also an area where several local development and planning initiatives dedicated to farmland protection and farming enterprises survival are currently carried out. So the level of conflict is quite high, and the local inhabitants are somewhat dissatisfied with decisions of building infrastructures of different types (highways, railroads, airport, windmills, new towns...) near the suburbs.

To start with, let us examine the geography of infrastructure-related conflicts[e] in the Greater Paris Region, the size of the pyramids indicating the number of legal proceedings. These conflicts are a good indicator of the process of peri-urbanization the region has been undergoing, a process that is met with much opposition from populations that reside in the areas concerned. Figure 2 shows that the infrastructure related conflicts are not randomly distributed in the Paris Region. On the contrary, they are concentrated in the area bordering the urban centre of Paris: One can see that the highly urbanized part of the Paris metropolis (Paris and its three bordering departments, or the *"petite couronne"* – little crown) seems little affected. The conflicts are revealing of the spatial constraints the Paris agglomeration faces in its process of expansion and in the construction of the infrastructures that are necessary to the implementation of new urban developments.

[e]They concern ,for the most part ,the projects of construction of road ,highway ,railway ,river or airport infrastructures ,as well aspubli cfacilitiessuchaswastewatertreatmentplants,townhalls,barracks,prisons ,multipurposehalls...

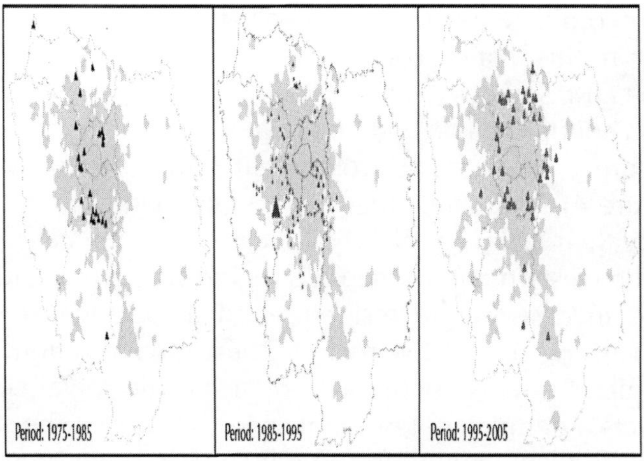

Figure 2: Geography of infrastructure-related conflicts in the Greater Paris Region from 1975 to 2005 (litigation court sources).

The evolution of infrastructure-related conflicts over three successive periods - 1975–1985, 1985–1995 and 1995–2005 - corresponds to the expansion of the areas represented in grey on the map, areas which correspond to municipalities with a population of over 5000 inhabitants. Over the last thirty years, the grey area has not expanded much but that the conflicts have multiplied in different places, all located at the border of the "*petite couronne*". They are peri-urban municipalities at the interface between the Paris agglomeration and the natural and agricultural areas which still represent over 50 percent of the total area of the Paris region. These municipalities have a high rate of urbanization (on average over 50 construction permits issued every year) and will no doubt become urban municipalities. The conflicts show that urban expansion does not always occur easily, because attempts to create infrastructures are confronted with organized opposition from residents who wish to preserve nature or their living environment. Conflicts here often take the form of opposition to decisions made by the public authorities, whether they concern the construction or the extension of infrastructure. They are a good expression of the process of legitimation of reflexivity that starts (Rosanvallon, 2008), in particular when it is based on logic of proximity with local populations exercising their democratic rights to intervene.

The *Conflicts* © database can also be used for examining the logics behind the location of conflicts such as they are revealed through the analysis of two different sources compared with the results of surveys conducted in this zone. For example, we observe heterogeneity between the DRP sources (with the daily newspaper «*Le Parisien*") and the litigation courts' data on conflicts related to infrastructures of public interest (Figure 3) or to urban development or the development of open spaces (Figure 4). The conflictual dynamics around the infrastructure identified by the local press clearly indicate that conflicts are far more intense in the western part of the Paris region where the value of land, the average income of the residents and associative activities are higher than in the rest of the region. Thus, mediatization, patent in the press articles, seems conditioned by a specific social context that gives conflicts a particular "territorial style". However, the similarities between the results obtained from both sources seem much stronger in the case of the conflicts related to urban development and the management of open spaces.

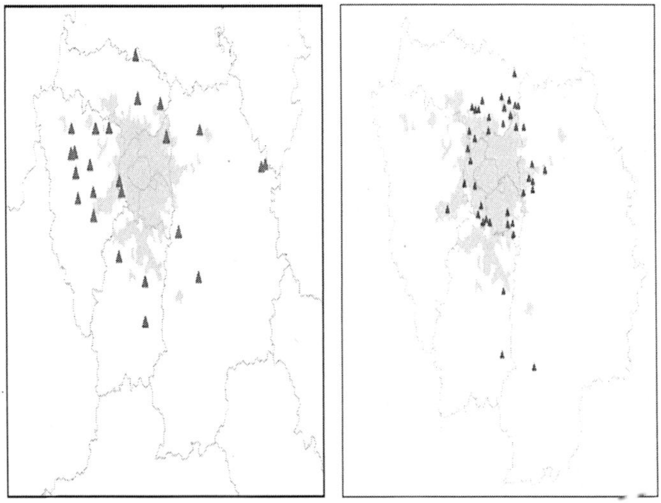

Figure 3: Conflicts related to infrastructure of public utility reported by the Daily Regional Press (2005) and litigation rulings (1999–2005) in the Greater Paris Region.

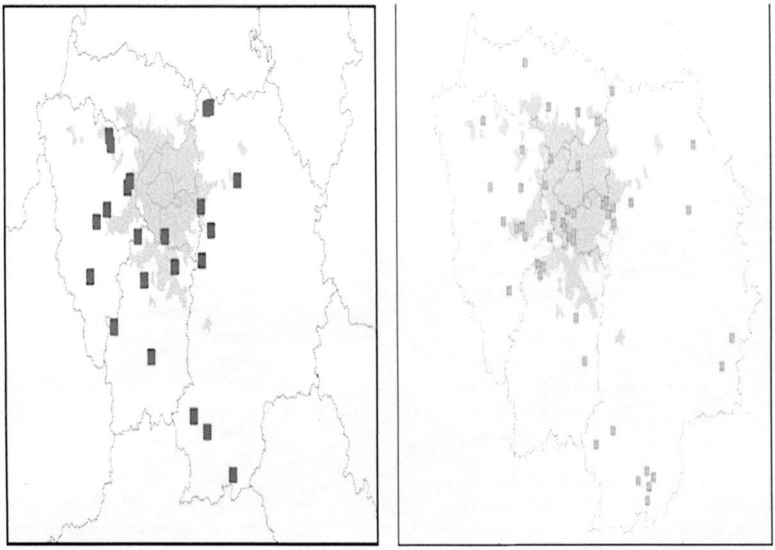

Figure 4: Conflicts related to urban development and the development of open spaces reported by the Daily Regional Press (2005) and litigation rulings (1999–2005) in the Greater Paris Region.

We can go further and undertake systematic comparisons between both sources, so as to compile complete and opposable profiles of conflictuality in the same zone. Figure 3, built from data covering the 2003–2005 period, provided by the DRP, highlights the different types of conflicts reported, as well as the number of press articles published about them. This gives us an idea of the intensity of those conflicts, as well as their respective repercussions. A distinction is made here between remedial conflicts (which start after the implementation of an infrastructure for example, or the occurrence of a contested action) and preventive conflicts (which are triggered when a project is publicized, for example, in the framework of a public enquiry pertaining to a building permit), a distinction that is possible thanks to the fact that the data have dates associated to them.

A process similar to that conducted with the DRP can be undertaken using the decisions made by tribunals, and more particularly by administrative tribunals. This leads to noticeably different results (Figure 5), which reveal the importance of using different sources of data for analyzing conflicts.

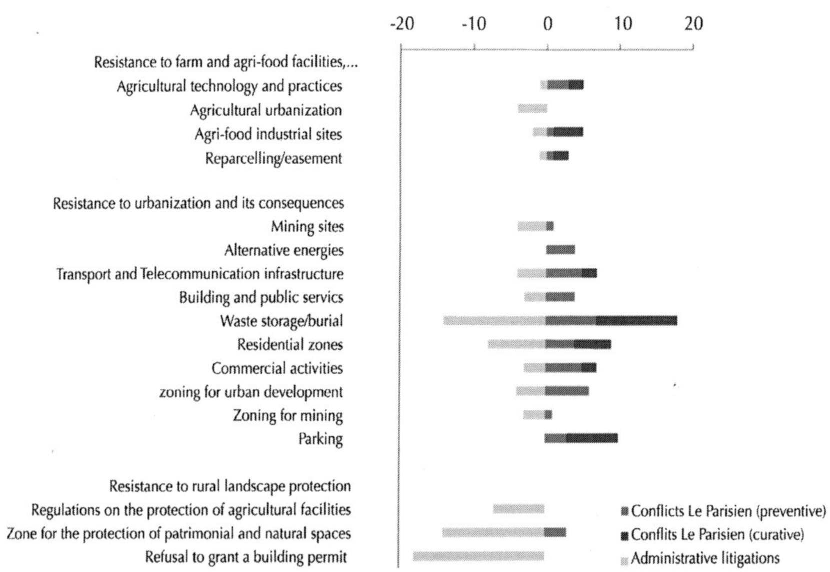

Figure 5: The main sources of land use conflicts in the Greater Paris Region, according to the administrative courts' rulings (Versailles, Cergy, Melun) (Source: *Le Parisien*, 2003–2005; Archive collections of Administrative tribu-

nals, 2005–2006); Darly2009.

The database can also be used for analyzing more precisely certain categories of conflicts - in this case the conflicts related (directly or indirectly) to the uses of agricultural land (Darly 2009; Darly and Torre 2013a, b). We make a distinction between three main groups of conflicts according to the characteristics of the contested facilities: the facilities dedicated to the functioning of the city – construction of transport or energy infrastructures, productive or residential activities, burial or spreading of waste, building and housing, commercial and industrial zones – the mechanisms related to agricultural economics in these territories - easements, re-parceling, and those related to landscaping projects – zones for the protection of land and natural resources, Regional Natural Parks – (Figure 6). Here again, we can see the importance of the conflicts related to the presence of the capital city (Paris) and to its expansion into peri-urban areas which regularly encroaches on land areas reserved for agricultural purposes or nature conservation and provokes the opposition of some of the people already residing in these areas.

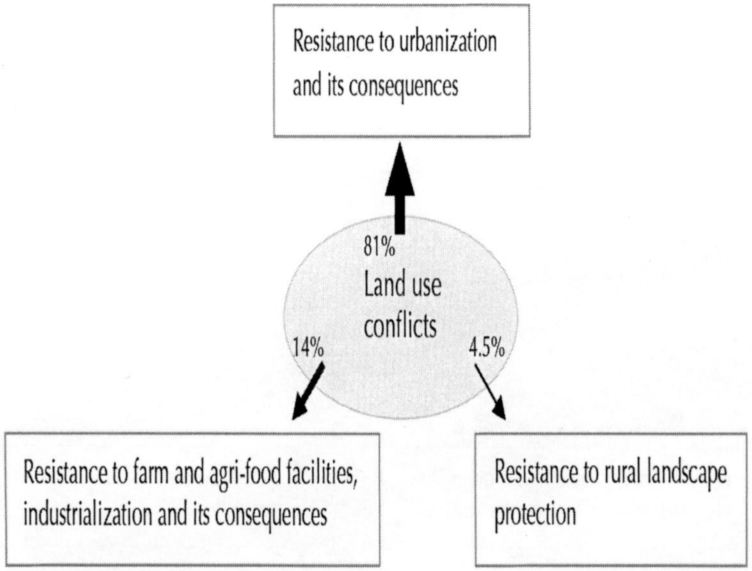

Figure 6: Conflict in the Greater Paris Region as observed by the DRP (Source: *Le Parisien*, 2003–205); Darly2009.

Comparing the results of the analysis of the DRP and of litigation ruling (Figure 7) shows noticeable differences in the case of conflicts related to the use of agricultural land: while the oppositions against the regulatory mechanisms for the protection of nature and open spaces represent a large number of litigation cases. The press reveals, more specifically, the collective and publicized dimension of the actions undertaken against so called urban sprawl interest projects (against infrastructures for example). This finding indicates that certain conflictual situations seem to lack the public dimension necessary for their publicization and therefore hardly feature in the press, but take the form of individual litigation cases. These cases mostly involve oppositions against administrative decisions such as land occupancy authorizations in urban law (building permits and planning certificates for example). Finally, a number of conflicts that have been the object of litigation concern similar categories of actors: It is the case of oppositions against some land restructuring projects, internal to the farming world. Inversely, it is mostly the confrontation between local residents and nonresident users that appears to be the main source of the conflicts that feature in the local press, conflicts in which the collective action dimension prevails.

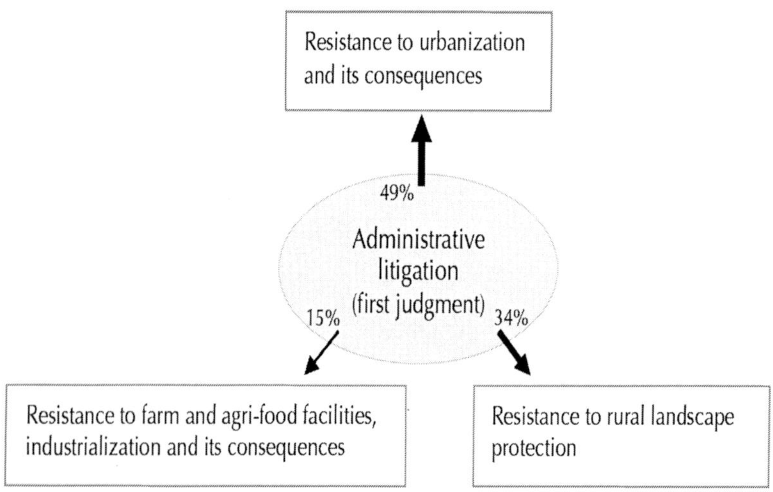

Figure 7: Conflict in the Greater Paris Region as described in legal cases dealt with by the Administrative Appeal Courts (Versailles and Paris) and the Council of State (1981–2005); Darly2009.

Conflicts in Developing Countries. an Example in South Pakistan

For a particular case study of land use conflict in developing countries, we have selected the case of Chotiari water reservoir from Pakistan; in order to put light on the land use conflicts caused by an infrastructural project setting with follow-up governance structure. This is one of Pakistan's largest infrastructure projects, which is facing opposition in the country and is held up as an example of weak governance in the planning of new infrastructures in developing countries. (Magsi and Torre, 2014) (see Figure 8).

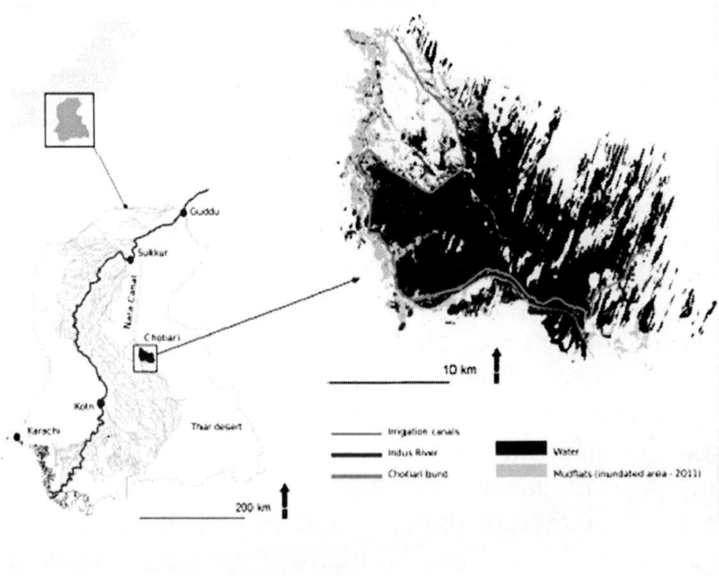

Figure 8: Location of the Chotiari water reservoir.

The Chotiari reservoir project was designed and implemented in order to increase the storage capacity of the existing lakes in the Chotiari wetlands and enable the irrigation of more arable land in Pakistan. The project was initiated in 1992 by the Water and Power Development Authority (WAPDA) and was funded by donor agencies via the World Bank. The project area extends over 18,000 hectares of entitled and unentitled land. The Chotiari reservoir area was characterized as wetlands and included lakes, forest, swamps, irrigation channels, agricultural land, barren land and a rich ecosystem, which supported the livelihoods of the local population through fishing, agriculture, grazing and a range of other economic activities. The reservoir project has created opposition between the principal actors (fishermen, farmers, livestock herders and others) on the one hand, and stakeholders from the public administration (national and provincial ministries), local politicians and landlords on the other. More specifically, a number of factors have made the task of implementing this project more complicated and controversial: the public administration's highly bureaucratic approaches and mismanagement of construction and compensation funds; local politicians' misuse of position and power with regard to forced displacements; and local landlords' exercise of power over the local population. Furthermore, opposition grew when local populations were dispossessed of their livelihoods and ancestral properties without proper compensation. In spite of all these issues, the public authorities completed and inaugurated the reservoir in February 2003, five years later than anticipated.

For this case study we have collected data through various sources, i.e. DRP, experts' opinion interviews by an open questionnaire and other sources (available literature, GIS and internet). The Chotiari project results reveal that local actors and outside stakeholders are under opposite aims and objectives of land use, and that the drivers of this situation (behaviors and interpersonal relations and actions) lead the project under situations of superposition of uses. This restlessness among stakeholders encouraged local journalists to demonstrate their issues. In fact, more than eighty percent of the articles (of total published news/articles in DRP since 1997–2011) reflected that there has been significant wrong doing associated with the land acquisitions, compensation and resettlement plans. Therefore, the project has brought on the dynamics of conflicts of land use, which has not only contributed in the changing of territoriality of the actors, but also came forth with disruption of socio-spatial practices. This interference of socio-spatial practices involved a reaction of discontent, which has sometimes expressed aggressively (Iqbal 2004). Moreover, public authorities have induced social and environmental nuisances by affecting arable lands, pasture, forest, as well as cruel displacement of local population. Besides all, the increasing water level in the reservoir is creating seepage[f], which is destroying adjacent agricultural lands (see Figure 9).

'The organizational management structure has allowed increasing water level in the reservoir for which purpose it has been constructed, but due to low standard earth work there is water seepage from embankments which is damaging agricultural lands outside the reservoir.

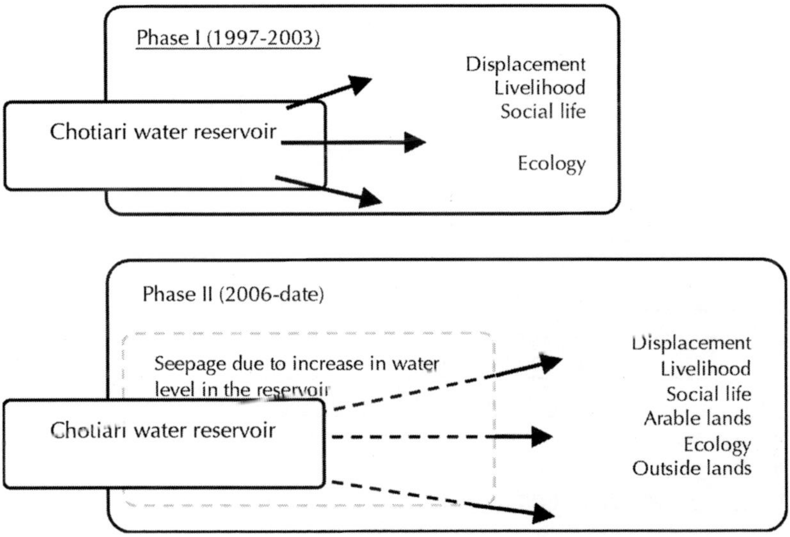

Figure 9: Conflict dynamics of Chotiari reservoir.

The multi-dimensional catastrophe of the Chotiari reservoir cannot be understood with a single factor. Therefore, it is important to visualize and quantify the structural and proximate factor dynamics with their anticipation, which have not only escalated conflicts of land use but also unrest among local population. Therefore, on the basis of articles published in daily press and opinions interviewed from experts, we came into effect to disclose the responsible factors to the conflicts of Chotiari reservoir (see Table 1). In this regard we have quantified the factors which appeared in DRP as well as in the expert opinion interviews. These factors seemed responsible for either giving favorable path to pre-conditions or conducive climate to the conflicts.

Table 1: Conflict factors of Chotiari reservoir

Factor types	Cause 0073	Percentage	
		DRP	Experts opinion
Structural factors	Corruption/misuse of funds	23.94	34.38
	Unilateral decision	21.81	21.88
	Lack of technical and scientific research	19.68	9.38
	International interest	7.98	12.50
	Non-existence of national resettlement policy	9.04	9.38
Proximate factors	Ethnic diversity and disarray (unrest among communities)	13.83	12.50
	Others (Nepotism, Illiteracy etc.)	3.72	0

Torree*tal.*

Torree*tal.SpringerPlus*20143:85, doi:10.1186/2193-1801-3-85

In the above table we see the differences in the factors of conflicts highlighted by either source, which may be due to the technical approaches and scientific analysis of both sources as well as the public character of DRP data compared with the privacy of face to face interviews. Therefore, we disclose the consequences (either positive or negative) of the project after deep analyzing of DRP as well as experts' opinion in the study area (see Figure10). After comparative analysis we found both similarities and dissimilarities in the data sources, due to privacy or the public character of their approaches.

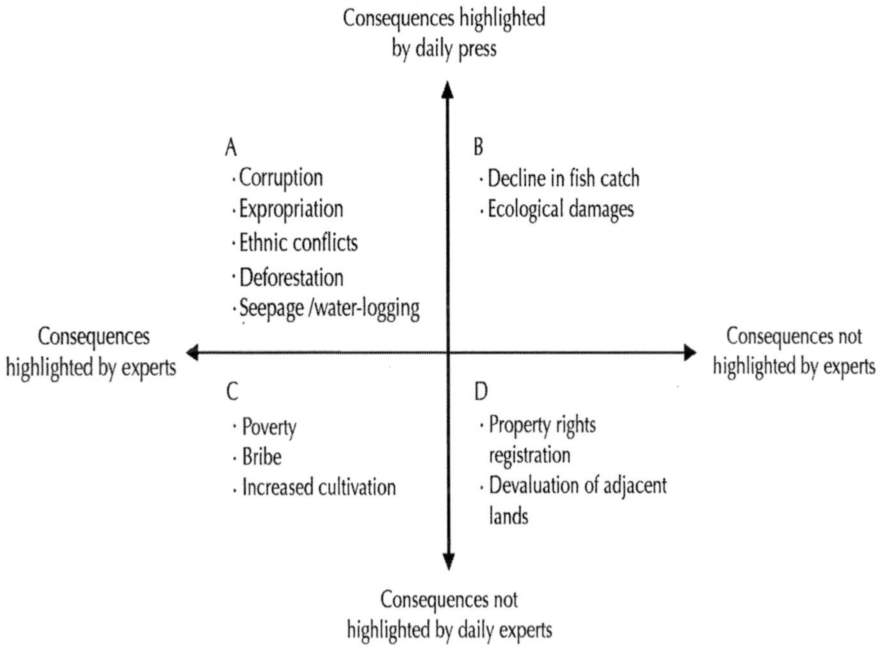

Figure 10: Consequences of Chotiari conflict: by daily press, experts and personal observations.

Results of this case study present the pattern of thought based on the expert opinions, daily press and interviews of affected households in the study area which insights a simple approach towards social representations in the categories of the actors. Thus it is difficult to categorize from planning to construction stages of the project. For example, during policy making process the actors were involved either from regional to national level, while some actors involved temporarily and did not play an active role in the administration. In order to understand the dynamic process of the project, we have tried to analyze a relational approach to provide some answers regarding social representations corresponding to a universe of interrelated elements. Though, after deep analyses we have summarized the relations, links and locations of the actors or stakeholders at different geographical scales (see Figure11).

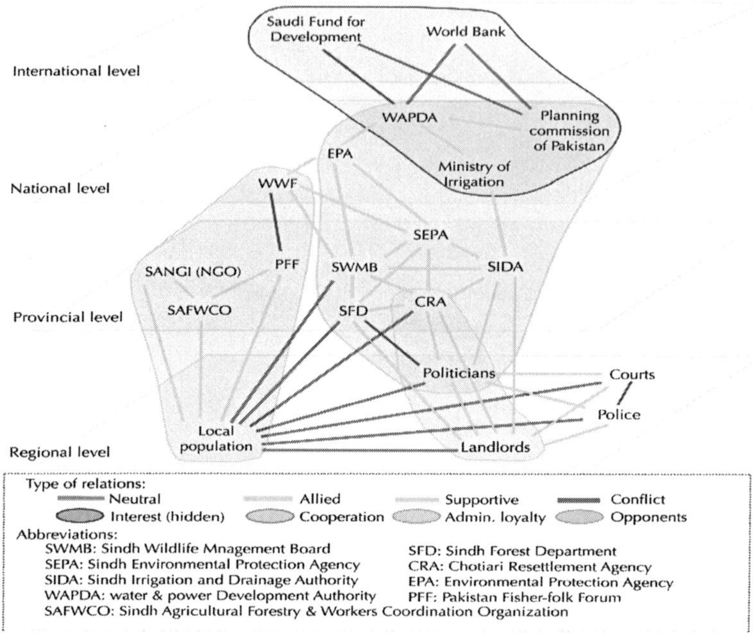

Figure 11: Networks of local actors in the Chotiari conflict.

Above figure exposes the network of the actors involved in Chotiari reservoir, with different relations and behaviors. The results show that those stakeholders have categorized themselves in defined strategies of construction and opposition of the reservoir (networks of pros and cons). Dramatically, the managing and administering actors (international to national scale) have a different representation on the reservoir area; they were found in alliance with the local politicians and landlords (regional scale), in order to construct the reservoir. According to the experts the cooperation of local stakeholders with administrative actors from international to national scales was based on some hidden interests of corruption and favoritism, while landlords and politicians have supported for construction reservoir. In this figure it is also disclosed that between two pro-construction networks (initiators and supporters), an administrative loyal network have played a key role to construct the reservoir. On the other hand, local actors were found in protestation against reservoir construction, relocation and compensation issues (Magsi and Torre, 2012). While, local population with the support of different NGOs have opposed decision of reservoir and manifested to protect natural resources of precious wetlands of Chotiari. Their limited scale of support led no valuable results of their oppositions, protests and agitations, because despite of this the reservoir has been constructed and inaugurated on February 2003. Results related to multilevel governance, we may say that this drama may be played to divert people's attention that conflicts are at regional scale only. Moreover, through above figure we have explored that law enforcement institutions (courts and police) are peripheral and seem more suppressive rather to have influence on the administration, in this situation the local population will surely have no hope of their violated rights. Through this case study we acquire the information on infrastructural development settings without agreement of local stakeholders. In fact, the only way to examine the institutional inconsistencies and distribution of dissimilar power, leading to land use conflicts and loss of local population's resources, is to analyze the dynamics of actors/stakeholders network in the study area, such as the reaction of all stakeholders during and after the project construction.

CONCLUSIONS

His article has aimed to present the work carried out in the last few years by a multidisciplinary team on the question of land use conflicts and to reveal the methodology of survey and data collection, as well as the structure of the resulting database. We have first presented the scope of our investigations by defining the conflicts in question, their characteristics, motives, manifestations and the actors involved. We have then presented our method of identification of conflicts, based on a diagnosis of the conflicts that occur in selected areas and on the combination of various methods of data collection, including interviews with experts, analyses of the DRP and of litigation rulings. Lastly, we have presented the *Conflicts* © database, with its tables and nomenclatures, which reconcile the data collected from different sources, before providing a few examples of how we have used implemented our method with cases of conflicts occurring in the Greater Paris Region.

Our studies, performed on the basis of the previous method, reveal that land use conflicts are characterized by a high diversity of expression depending on the activities, on the uses around which they emerge, on the territories in which they occur, as well as on the characteristics of the actors involved in the conflicts in question. Furthermore, some conflicts, which are closely related to certain specialized activities, are kept relatively private and may even be limited to face to face interactions between two private parties, whereas others, related to decisions concerning a large number of individuals and involving questions of land use regulations, are liable to involve the participation of the public authorities and associations representing part of the population (for example conflicts related to the definition of local urban development plans or of protected zones). The quantitative survey, crossing various sources of information, and the qualitative surveys show that land use conflicts can take extremely diverse forms of expression and manifestations, but are however centered on large categories of conflicts, territories or modes of resistance to undesirable projects. Thus, observing conflictuality has nothing to do with the mere collection of raw information expressing a reality that is easy to decrypt. The goal of the crossed analysis of various information materials is precisely to explain the fact that the modes of expression of conflicts do not just constitute «a source of information» but a framework of observation that determine the types of phenomena observed.

Our research into conflicts in rural and periurban areas shows that this dimension is key in processes of territorial management, regional development or the governance of various local activities. Sometimes, conflicts are blind oppositions or are the product of egoistical Nimby behaviours. But in many cases they constitute a way of initiating discussions on the issues and paths of territorial development and of influencing decisions by participating in processes underway from which one had been excluded. That is why they have a bearing, either on the decisions on land use and management (arbitrated negotiation) or on the composition and representativeness of the bodies responsible for taking decisions (arbitral negotiation). The conflict thus becomes an integral part of the deliberative process at the local level by allowing an expression of local democracy and the re-inclusion of participants who were forgotten or deliberately excluded during earlier project development stages.

Land-use conflicts thus constitute one form of resistance and expression of opposition to decisions that leave part of the local population unsatisfied. Some local innovations, whether technical or organizational in nature, give rise to resistance which can turn into conflict. Major changes requiring reconfiguration of the use of space (creation of transport, energy or waste-processing infrastructure, new urban master plans, territorial or environmental zoning, etc.) generate conflicts whose spatial and social extent can quickly grow. Conflicts are signals of social, technological and economic changes, indicators of novelty and innovations. They demonstrate the opposition aroused by the latter, lead to discussions on their implementations and their possible (non-) acceptability as well as on the adoption of governance procedures and their transformation under the influence of the dynamics of change. All changes encounter opposition or resistance of varying relevance and justification. But it would however be simplistic to see this resistance as a systemic sign of reactionary opposition to change because, in a number of cases, they are more a reflection of differences over the direction taken by the new initiatives that are being imposed on the public than of a stubborn desire to maintain the status quo. During these phases of conflict, social and interest groups tend to reconstitute themselves and may even undergo technical or legal changes. Once a conflict ends, it leaves behind new local agreements, new modes of governance, new configurations of discussion forums as well as new technical procedures (changes in direction, various adjustments, changes in urban planning documents, etc.), all arrived at during the negotiations.

Territorial governance processes are today undergoing intense upheavals and are subject to intense periods of discussions and conflict oppositions. These latter shape the phases of territorial innovation and thus change the directions of development and growth in rural or urban territories. Such governance mechanisms and their conflict sides can be viewed as laboratories of change because they accompany and sometimes anticipate the changes underway in the territories by giving them shape, by helping maintain a dialogue and expressions of opposition and by preventing violent confrontations or failures of development due to sluggishness or expatriation. Therefore, these changes in land use occupations and the subsequent oppositions they gave birth to are embodied in the opposing and twin forms of conflict and consultation which constitute the modes of expression and the vehicles of transmission of on-going innovations at the territorial level.

AUTHORS'CONTRIBUTIONS

AT inspired and directed the program, and wrote most of the article. RM greatly participated to the elaboration of the data collection method and to the data base elaboration, and wrote most of the jurisdictional parts of the article. HM elaborated the method for developing countries and wrote most of the developing countries example section (Chotiari, Pakistan). LB participated to the elaboration of the data collection method and to the data base elaboration, and performed several case studies. AC participated to the elaboration of the data collection method and to the data base elaboration, and performed several case studies. AC participated to the elaboration of the data collection method and to the data base elaboration, and performed several case studies. SD participated to the elaboration of the data collection method and to the data base elaboration, and wrote most of the developed countries example section (Ile-de-France, France). PhJ participated to the elaboration of the data collection method and to the data base elaboration, and performed several case studies. TK inspired and directed the program, greatly participated to the elaboration of the data collection method and to the data base elaboration, and performed several case studies. HVP participated to the elaboration of the data collection method and to the data base elaboration, and to the statistical study of litigation issues. OK participated in the elaboration of the method for developing countries and performed several case studies. All authors read and approved the final manuscript.

ACKNOWLEDGEMENTS

Some published outputs of the research program: Anthopoulou T, Moissidis A (2002) La périurbanisation dans l'espace rural grec: crise ou adaptation? Géocarrefour 7(4):359–366; Bonin M, Torre A (2004) Typologie de liens à l'espace impliqués dans les conflits d'usage, Etude de cas dans les Monts d'Ardèche. Les Cahiers de la multifonctionnalité 5:17–31; Bossuet L, Torre A (2009) Le devenir des ruralités, entre conflits et nouvelles alliances autour des patrimoines locaux. Economie Rurale 313–314:147–162;; Cadoret A (2006) Conflits d'usage liés à l'environnement et réseaux sociaux : enjeux d'une gestion intégrés ? Le cas du littoral du Languedoc-Roussillon. PhD thesis of Geography.

Université de Montpellier 3, France; Caron A, Torre A (2002) Les conflits d'usage dans les espaces ruraux. Une analyse économique. In: Perrier-Cornet P (ed) A qui appartient l'espace rural? Editions de l'Aube, Paris; Caron A, Torre A (2006) Vers une analyse des dimensions négatives de la proximité. In: Les conflits d'usage et de voisinage dans les espaces naturels et ruraux. Développement Durable et Territoires, n7, online; Darly S (2008a) La spatialité des conflits d'usage au sein des zones périurbaines en Ile-de-France: analyse empirique d'une modalité peu connue de la gouvernance des territoires. Norois 209(4):127–146; Darly S (2008b) Tensions et conflits d'usage liés à l'agriculture. Géographie de deux corpus d'observation au sein de la région Ile-de-France. In: Loudiyi S, Bryant CR, Laurens L (ed) Territoires périurbains et Gouvernance. Perspectives de recherche, Université de Montréal, mai, pp 109–117; Darly S (2008c) Conflits d'usage et aires conflictuelles à l'échelle d'une région métropolitaine. In: Gorgeon C, Laudier I (ed) Le cas des enjeux liés à l'espace agricole en Ile-de-France. Territoires et identités en mutation, l'Harmattan, Paris, pp 87–106; Darly S, Torre A (2008) Conflits liés aux espaces agricoles et périmètres de gouvernance en Ile-de-France (résultats à partir d'analyses de la presse quotidienne régionale et d'enquêtes de terrain). Geocarrefour 83(4):307–319; Jeanneaux P, Perrier-Cornet P (2008) Les conflits d'usage du cadre de vie dans les espaces ruraux et la décision publique locale, Éléments pour une analyse économique. Economie rurale 306:39–54; Jeanneaux P, Sabau C (2009) Conflits environnementaux et décisions juridictionnelles : que nous apprend l'analyse du contentieux judiciaire dans un département français ? VertigO - la revue électronique en sciences de l'environnement 9:1. URL:http://vertigo.revues.org/index8412.html; Kirat T, Melot R (2006) Du réalisme dans l'analyse des conflits d'usage: les enseignements de l'étude du contentieux. Développement durable et territoire. Revue numérique; Kirat T, Torre A (ed) (2008) Territoires de Conflits. Analyses des mutations de l'occupation de l'espace, l'Harmattan, Paris; Kirat T, Torre A (2007) Quelques points de repères pour évaluer l'analyse des conflits dans les théories économiques, avec une emphase particulière sur la question spatiale. Géographie, Economie, Société 9(2):215–240; Kirat T, Torre A (ed) (2006 & 2007) Conflits d'usages et dynamiques spatiales les antagonismes dans l'occupation des espaces périurbains et ruraux (I & II). Géographie, Economie, Société, *8*(3) & *9*(2); Lefranc C, Torre A (2004) Tensions, conflits et processus de gouvernance dans les espaces ruraux et périurbains français. In: Scarwell HJ, Franchomme M (ed) Contraintes environnementales et gouvernance des territoires. Eds de l'Aube, p 469; Magsi H (2013) Land use conflicts in developing countries, framing conflict resolution and prevention strategies to ensure economic growth and human welfare. The case of Chotiari water reservoir from Pakistan. PhD thesis of Economics. AgroParisTech, Paris, 209p; Magsi H, Torre A (2013) Approaches to understand land use conflicts in the developing countries. The Macrotheme Review 2(1):119–136; Paoli J-C, Melot R, Fiori A (2008) L'aménagement du territoire à l'épreuve de la décentralisation : conflits et concertation en Corse et Sardaigne. Pôle Sud: revue de science politique de l›Europe méditerranéenne 28(1):143–165; Torre A (2008) Conflits d'usage dans les espaces ruraux et périurbains. In: Monteventi WL, Deschenaux C, Tranda-Pittion N (ed) Campagne-ville. Le pas de deux, Presses Polytechniques et Universitaires Romandes, Lausanne; Torre A, Aznar O, Bonin M, Caron A, Chia E, Galman M, Guérin M, Jeanneaux P, Kirat T, Lefranc C, Melot R, Paoli JC, Salazar MI, Thinon P (2006) Conflits et tensions autour des usages de l'espace dans les territoires ruraux et périurbains, Le cas de six zones géographiques françaises. Revue d'Economie Régionale et Urbaine 3:415–453; Torre A, Caron A (2005a) Réflexions sur les dimensions négatives de la proximité : le cas des conflits d'usage et de voisinage. Economie et Institutions 6 & 7:183–220; Torre A, Caron A (2005b) Conflits d'usage et de voisinage dans les espaces ruraux. In: Torre A, Filippi M (ed) Proximités et Changements Socio-économiques dans les Mondes Ruraux. INRA éditions, Paris, p 322; Torre A, Lefranc C (2006) Les Conflits dans les zones rurales et périurbaines, Premières analyses de la Presse Quotidienne Régionale. Espaces et Sociétés 124–125(1–2):93–110; Torre A, Melot M, Bossuet L, Cadoret A, Caron A, Darly S, Jeanneaux P, Kirat T, Pham HV (2010) Comment évaluer et mesurer la conflictualité liée aux usages de l'espace ? Eléments de méthode et de repérage. VertigO - la revue électronique en sciences de l›environnement, Vol. 10, N° 1.http://vertigo.revues.org/9590.

REFERENCES

1. Azuela A, Herrera-Martin C (2010) Taking Land around the World: International Trends in Expropriation for Urban and Infrastructure

Projects. In: Lall SV, Freire M, Yuen B, Rajack R, Helluin JJ (eds) Urban Land Markets, Improving Land Management for Successful Urbanization, Heidelberg: Springer Verlag.

2. Barré M-D, Aubusson de Cavarlay B, Zimolag M (2006) Dynamique du contentieux administratif. Analyse statistique de la demande enregistrée par les tribunaux administratifs. Ministère de la Justice, Paris: Rapport pour la Mission de recherche Droit et justice.
3. Boulding KE (1962) Conflict and Defense. New York: Harper and Row.
4. Cadene P (1990) L'usage des espaces périurbains. Une géographie régionale des conflits. Economie Rurale 118–119:235-267
5. Cadoret A (2009) Conflict dynamics in coastal zones: a perspective using the example of Languedoc-Roussillon (France. Journal of Coastal Conservation: planning and management 13:151-163
6. Campbell DJ, Gichohi H, Mwangi A, Chege L (2000) Land use conflict in Kajiado District, Kenya. Land Use Policy 17(4):337-348
7. Castro AP, Nielsen E (2001) Indigenous people and co-management: implications for conflict management. Environ Sci Policy 4(4):229-239
8. Charlier B (1999) La défense de l'environnement : entre espace et territoire, géographie des conflits environnementaux déclenchés en France depuis 1974. France: PhD thesis, Université de Pau et des Pays de l'Adour. p 750
9. Coser LA (1982) Les fonctions du conflit social. Paris: PUF.
10. Darly S (2009) Faire coexister ville et agriculture au sein des territoires périurbains. Antagonismes localisés et dynamiques régionales de la conflictualité. PhD Thesis of Ecole des Hautes Etudes en Sciences Sociales, Paris 457:p. + Annexes
11. Darly S, Torre A (2013) Conflicts over farmland uses and the dynamics of "agri-urban" localities in the greater Paris region. Land Use Policy 33:90-99
12. Darly S, Torre A (2013) Land-use conflicts and the sharing of resources between urban and agricultural activities in the Greater Paris Region. Results based on information provided by the daily regional press. In: Noronha-Vaz T, Leeuwen VE, Nijkamp P (eds) Towns in a rural world, London: Ashgate.

13. Deininger K, Castagnini R (2006) Incidence and impact of land conflict in Uganda. J Econ Behav Organ 60:321-345
14. Diehl P (1991) Geography and War: A Review and Assessment of the Empirical Literature. Int Interact 17:11-27
15. Dziedzicki JM (2001) Gestion des conflits d'aménagement de l'espace : Quelle place pour les processus de médiation?. Thèse pour le doctorat d'aménagement de l'espace et urbanisme: Université de Tours, Tours.
16. Freund J (1983) Sociologie du Conflit. Paris: PUF.
17. Fisher R (1997) Interactive conflict resolution. Syracuse, New York: Syracuse University Press.
18. Galman M, the participants of Conflict Program (2007) Guide Base de données Conflits. UMR SAD-APT. p 64
19. Hensel PR (2001) Contentious issues and world politics: the management of territorial claims in the Americas, 1816–1992. Int Stud Q 45:81-109
20. Humphreys M (2005) Natural resources, conflict, and conflict resolution: uncovering the mechanisms. J Confl Resolution 49(4):508-537
21. Iqbal N (2004) Affectees of Tarbela and Chotiari Dams: A struggle for social justice. Addressing Existing Dams: United Nations Environmental Program. pp 69-72 Issue based workshop
22. Janelle D (1977) Structural Dimensions in the Geography of Locational Conflicts. Can Geogr 21:311-328
23. Jeong HW (1999) Conflict management and resolution. In: Kurtz L (ed) Encyclopedia of Violence, Peace and Conflict, vol 1. Academic Press, pp 389-400
24. Joerin F, Pelletier M, Trudelle C, Villeneuve P (2005) Analyse spatiale des conflits urbains, Enjeux et contextes dans la région de Québec, Cahiers de Géographie du Québec. Numéro thématique Conflits, Proximité, Coopération 49(138):319-342
25. Kirat T, Torre A (2004) Modalités d'émergence et procédures de résolution des conflits d'usage autour de l'espace et des ressources naturelles. In: Analyse dans les espaces ruraux. vie et société: Chapter 2. Rapport de recherche CNRS, Programme Environnement.

26. Lascoumes P (1995) Les arbitrages publics des intérêts légitimes en matière d'environnement, L'exemple des lois Montagne et Littoral. Revue française de science politique 45(5):396-419
27. Leost P (1998) La stratégie contentieuse d'une association de protection de la nature en Bretagne. In: Le Louarn P (ed) Décision locale et droit de l'environnement, Etude comparée des cas breton et martiniquais. Rennes: Presses Universitaires de Rennes. pp 85-106
28. Leeuwen M (2010) Crisis or continuity? Framing land disputes and local conflict resolution in Burundi. Land Use Policy 27(3):753-762
29. Lewin K (1948) Resolving Social Conflicts. New York: Harper & Row.
30. Ley D, Mercer J (1980) Locational conflicts and the politics of consumption. Econ Geogr 56(2):89-109
31. Magsi H, Torre A (2014) Proximity analysis of inefficient practices and socio-spatial negligence: Evidence, evaluations and recommendations drawn from the construction of Chotiari reservoir in Pakistan. Land Use Policy 36:567-576
32. Magsi H, Torre A (2012) Social network legitimacy and Property right loopholes: Evidences from an infrastructural water project in Pakistan. J Infrastructure Dev 4(2):59-76
33. Mann C, Jeanneaux P (2009) Two Approaches for Understanding Land-Use Conflict to improve Rural Planning and Management. J Rural Community Dev 4(1):118-141
34. Melé P, Larrue C, Rosenberg M (2004) Conflits et territoires, Presses Universitaires François Rabelais. Tours: collection perspectives « Villes et territoires ».
35. Mc-Carthy JD, Mc-Phail C, Smith J (1996) Images of protest: dimensions of selection bias in media coverage of Washington demonstrations, 1982–1991. Am Sociol Rev 39:101-112
36. Neslund N (1990) Dispute resolution: a matrix of mediation. J Dispute Resolution 2:217-266
37. Observatorio Permanente dos Conflitos Urbanos na Cidade de Rio de Janeiro (2010) http://www.observaconflitos.ippur.ufrj.br/novo/ajax/relatal.asp

38. Olzak S (1992) The dynamics of ethnic competition and conflicts. Stanford: Stanford University Press.

39. Owen L, Howard W, Waldron M (2000) Conflicts over farming practices in Canada: the role of interactive conflict resolution approaches. J Rural Stud 16:475-483

40. Pham HV, Kirat T (2008) Les conflits d'usage des espaces périurbains et le contentieux administratif - Le cas de la Région Ile-de-France. Revue d'Economie Rurale et Urbaine 5:671-700

41. Pham HV, Kirat TH, Torre A (2012) Les conflits d'usage dans les espaces ruraux et périurbains. Le cas des infrastructures franciliennes. Economie Rurale 332:9-30

42. Rosanvallon P (2008) La légitimité démocratique. Paris: Le Seuil.

43. Rucht D, Neidhardt F (1999) Methodological Issues in Collecting Protest Event Data: Unit of Analysis, Sources and Sampling, Coding Problems. In: Rucht D, Koopmans R, Neidhardt F (eds) Acts of Dissent: New Developments in the Study of Protest, Lanham: Rowman and Littlefield Publishers. pp 65-89

44. Simmel G (2008) Sociological Theory. New York: McGraw–Hill.

45. Starr H (2005) Territory, Proximity, and Spatiality: The Geography of International Conflict. Int Stud Rev 7:387-406

46. Stephenson GM (1981) Intergroup bargaining and negotiations. In: Turner JC, Giles H (eds) Intergroup Behaviour, Oxford: Basil Blackwell.

47. Struillou J-F (2004) L'application du droit pénal de l'urbanisme, In Etat de droit et urbanisme. Cahiers du Gridauh 11:87-126

48. Touraine A (1978) La voix et le regard. Paris: Seuil.

49. Trudelle C (2003) Au-delà des mouvements sociaux : une typologie relationnelle des conflits urbains. Cahiers de Géographie du Québec 47(131):223-242

50. Wieviorka M (2005) La violence. Paris: Hachette Littératures.

Citations

CHAPTER 1

Matthew J. Bell, Brendan R. Cullen, and Richard J. Eckard, The Influence of Climate, Soil and Pasture Type on Productivity and Greenhouse Gas Emissions Intensity of Modeled Beef Cow-Calf Grazing Systems in Southern Australia, doi:10.3390/ani2040540.

CHAPTER 2

Gerhard Flachowsky[1] and Josef Kamphues[2], Carbon Footprints for Food of Animal Origin: What are the Most Preferable Criteria to Measure Animal Yields?, doi:10.3390/ani2020108.

CHAPTER 3

Naomi J. Fox , Glenn Marion, Ross S. Davidson , Piran C. L. White, and Michael R. Hutchings, Livestock Helminths in a Changing Climate: Approaches and Restrictions to Meaningful Predictions, doi:10.3390/ani2010093.

CHAPTER 4

Seifu Hagos, Torleif Lunde, Damen H Mariam, Tassew Woldehanna, and Bernt Lindtjørn, Climate Change, Crop Production and Child Under Nutrition in Ethiopia; a Longitudinal Panel Study, doi:10.1186/1471-2458-14-884.

CHAPTER 5

Yingcheng Liu, Tao Zhang, Xiaoli Geng, Liansheng He, and Zhiguo Pang, "Herdsmen's Adaptation to Climate Changes and Subsequent Impacts in the Ecologically Fragile Zone, China," Advances in Meteorology, vol. 2013, Article ID 748715, 8 pages, 2013. doi:10.1155/2013/748715.

CHAPTER 6

Christian S. Otchia, Agricultural Modernization, Structural Change and Pro-poor Growth: Policy Options for the Democratic Republic of Congo, doi: 10.1186/s40008-014-0008-x.

CHAPTER 7

Ewuketu Linger, Agro-Ecosystem and Socio-Economic Role of Homegarden Agroforestry in Jabithenan District, North-Western Ethiopia: Implication for Climate Change Adaptation, doi:10.1186/2193-1801-3-154.

CHAPTER 8

André Torre, Romain Melot, Habibullah Magsi, Luc Bossuet, Anne Cadoret, Armelle Caron, Ségolène Darly, Philippe Jeanneaux, Thierry Kirat, Haï Vu Pham, and Orestes Kolokouris, Identifying and measuring land-use and proximity conflicts: methods and identification,. doi:10.1186/2193-1801-3-85.

Index